A Third Window

A Third Window

Natural Life beyond Newton and Darwin

Robert E. Ulanowicz

Templeton Foundation Press

West Conshohocken, Pennsylvania

Templeton Foundation Press
300 Conshohocken State Road, Suite 550
West Conshohocken, PA 19428
www.templetonpress.org

Designed and typeset by Kachergis Book Design

LIBRARY OF CONGRESS CATALOGING-IN-PUBLICATION DATA
Ulanowicz, Robert E.
A third window : natural life beyond Newton and Darwin / Robert E. Ulanowicz.
p. cm.
Includes bibliographical references and index.
ISBN-13: 978-1-59947-154-9 (pbk. : alk. paper)
ISBN-10: 1-59947-154-x (pbk. : alk. paper)   1. Ecology—Philosophy.
2. Bateson, Gregory, 1904–1980.   I. Title.
QH540.5.U438 2009
577.01—dc22   2008040963

Printed in the United States of America
09 10 11 12 13 14   10 9 8 7 6 5 4 3 2 1

Tables 3.1 and 3.2 reprinted from Ulanowicz, R. E. 1999. Life after Newton: An
ecological metaphysic. *Biological Systems* 50: 127–42, with permission of Elsevier.
Figure 3.1, "Pedestrians—The Airport," reprinted with permission from James
Zwadlo, Milwaukee, WI.
Figures 3.2 and 3.3 reprinted from Ulanowicz, R. E. 2007. Emergence, naturally!
Zygon 42 (4): 945–60, with permission of Wiley-Blackwell Publishing.
Figures 4.1, 4.2, 4.3, 4.4, 4.5, and 5.2 reprinted from Ulanowicz, R. E. 1997. *Ecology,
the Ascendent Perspective.* New York: Columbia University Press, with permission.
Figure 4.6 reprinted from Ulanowicz, R. E., Goerner, S. J., Lietaer, B., Gomez, R. In
press. Quantifying sustainability: Resilience, efficiency and the return of informa-
tion Theory. *Ecological Complexity.*
Figure 5.1 reprinted from Ulanowicz, R. E. 2004. New perspectives through brackish water ecology. *Hydrobiologia* 514: 3–12, with permission of Springer Science +
Business Media.
Figure 5.3 reprinted from Ulanowicz, R. E. 1983. Identifying the structure of cycling
in ecosystems. *Mathematical Biosciences* 65: 219–37, with permission of Elsevier.

For Anya, Peter, and Vera, with fond memories of Little Bear, Baby Alligator, and Wild Pig.

Contents

Foreword
The Open Universe

Robert Ulanowicz has written a deeply important, controversial, and potentially transformative book. My aim in this foreword is not to speak for Ulanowicz, but to briefly outline his central claims and then discuss a broad context in which his views, and my own, discussed below, fit. At stake, in my view, may be the need for a radical post-reductionist science to complement and perhaps augment reductionism.

At its core, *A Third Window* seeks to go beyond both reductionism, the first window on the world, captured in the Newtonian worldview with its time reversible laws; and Darwin, who brought history deeply into the second window on the world, with a third window based on process ecology. Ulanowicz makes major claims. First, with the philosopher Karl Popper, he wishes to relax the concept of strict causality to Popper's more general idea of "propensities" and to suggest that in the biological realm propensities are a more realistic description of the world than any firmer "causality." The most radical aspect of the third window is that there are "causal holes" in the fabric of space/time. In place of causality, Ulanowicz argues for *raw chance*, the aleatoric. He bases this radical claim on two previous sources, Bertrand Russell and Alfred North Whitehead, and one of Niels Bohr's last students, Walter Elsasser. Russell and Whitehead had claimed that natural law must be based on homogeneous classes, such as the set of all identical electrons. Elsasser argued for the unique heterogeneous combinatorics of organisms, where by unique and heterogeneous, Elsasser meant heterogeneous features of organisms that could reasonably occur only once in the history of the universe. Ulanowicz wishes to say that in such circumstances, causality is not applicable, but propensities are applicable. The powerful consequence of this lack of causality is a *lack of natural law* capable of describing the aleatoric unique combinatorial events which arise. Thus, the most radical claim, the cornerstone of the third window onto the world, is that the unfolding of the universe is not entirely describable by natural law. The final central point of *A Third Window* is based on the general idea of "autocatalysis" or mutualisms, in which we replace a focus on objects as the center of our attention and focus instead on processes. An autocatalytic set of interwoven processes is one in which, in the simplest case, a process A abets process B, which in turn aids process A. More generally, a rich web of processes can be collectively "autocatalytic" or mutualistic. Such a set of processes can evolve from the top down, in which A is replaced by A′, a new process which helps B better than did A. Here causality is top down, rather than bottom up, as reductionists would hold. A′ replaces A in the mutualistic cycle because the entire cycle functions more efficiently with A′ than A, hence is selected by Darwin's natural selection.

The third window, in this brief description, opens a view of the biotic world beyond the reach of sufficient natural law, where causality fails in the face of unique combinatorial diversity, the aleatoric, and where top-down organization of autocatalytic systems of linked processes under selection is what drives the evolution of ecosystems and the biosphere quite as much as bottom-up mutations. This third window is, then, a radical new view of the biotic world.

What I should like to do now is attempt to place the bold effort by Ulanowicz in a broad framework that is strongly supportive of the third window, even though I am not yet convinced of the raw chance, the aleatoric that Elsasser and Ulanowicz argue for. To do this, and with prior discussion with the author, I want to put the issues in the framework of what I will call "the open universe." Like Ulanowicz, my most radical claim will be that the unfolding of the universe is *not sufficiently describable by natural law,* a claim I have discussed in two books, *Investigations* and *Reinventing the Sacred.*

Consider Pierre Laplace and his famous demon, an intelligence which, if given the positions and momenta of all the particles in the universe could, using Newton's time reversible laws, compute the entire future and past of the universe. This is perhaps the simplest statement of reductionism. If we add fields, including quantum field theory, the standard model, and general relativity we have, in outline, modern physics and contemporary reductionism where Nobel laureate Stephen Weinberg claims that all the explanatory arrows point downward from societies to people, to organs, to cells, to biochemistry, to chemistry, and finally to physics. In a recent communication, Weinberg told me that he did not care about the capacity of physical laws to predict all in the universe, rather he cared that all that happened in the universe was "entailed" by the laws of physics.

There are a number of features of Laplace's reductionism worth stressing: 1) The universe is deterministic—thrown into doubt a century later by quantum mechanics, the standard Copenhagen interpretation and Born's rule. 2) The only things that are ontologically real in the universe are "nothing but" particles in motion. A man found guilty of murder is nothing but particles in motion. 3) All that unfolds in the universe is describable by natural law. 4) There exists at least one language which is sufficient to describe all of reality—here Newton's laws and atoms in the void. 5) There are no causal holes in the fabric of space/time.

I believe that 1, 2, and 3 above are wrong and am open to the failure of 4 and 5.

As I discuss in *Reinventing the Sacred*, even physicists such as Nobel Laureates Philip Anderson and Robert Laughlin doubt the adequacy of reductionism and now argue for emergence. More, I think biology is not reducible to physics. Grant that Weinberg, *given* all the properties of your heart, could deduce all its properties from the laws of physics, he would have no way to answer Darwin's point that the function of the heart is to pump blood and that the *heart came into existence in the universe* as a complex organ and set of processes precisely because it pumped blood. Weinberg could deduce, in principle, all the properties of the heart, but not pick out pumping blood as particularly relevant. But Darwin would tell us that the heart was selected to pump blood. I claim that Weinberg cannot deduce or simulate the coming into existence of the heart in the universe. Nor is it obvious in what sense, if any, is the coming into existence in the universe of the heart "entailed" by the laws of physics.

I now take a step somewhat similar to Elsasser's and Ulanowicz's with respect to their unique heterogeneous events. Consider all proteins of length 200 amino acids. There are 20 to the 200th power or 10 to the 260th power such proteins. Were the 10 to the 80th particles in the universe to do nothing but make proteins length 200 on the Planck time scale, it would require 10 to the 39th times the lifetime of the universe to make all these proteins just once. Thus, the unfolding of the universe above the level of atoms is grossly nonrepeating, or nonergodic. The universe is on a unique trajectory with respect to possible complex molecules, organisms, or social systems, and *indefinitely open "upward" in complexity*. History enters the universe when the space of the possible is much larger than the space of the actual.

Next, let us consider what are called Darwinian preadaptations. Darwin noted that a feature of an organism of no use in the current selective environment might become of use in some different environment so be selected, typically for a novel functionality. I give one example. Swim bladders occur in certain fish and the level of water and air adjusts neutral bouyancy in the water column. Paleontologists claim that swim bladders evolved from lung fish. Water got into the lungs of some fish, creating a sac with air and water which was poised for a novel use as a swim bladder. Selection then selected for this novel functionality in the biosphere. Now obviously such a new function emerged in the biosphere. Critically, the new functionality had cascading consequences in the further evolution of the biosphere with new species and new proteins and other molecules. I now come to my central question. Can we say *ahead of time* all possible Darwinian preadaptations of all organisms alive now, or just for humans? That answer seems to be a clear "no". We seem entirely unable to prestate finitely all possible Darwinian preadaptations for humans or any other evolving organism. Part of the problem seems to be these: How would we prestate the selective conditions leading to the preadaptation being selected for the new functionality? And how would we prespecify the aspects of one or several organisms that might constitute the preadaptation so selected? Yet such preadaptations occur all the time in the evolution of the biosphere. Let me introduce the idea of the "adjacent possible" of the biosphere. Once there were lung fish, the swim bladder was in the adjacent possible of the biosphere. When there were no multi-celled organisms, the swim bladder was not in the adjacent possible of the biosphere. Then what appears to be true is that we cannot prestate the adjacent possible of the biosphere.

Very powerful consequences follow from this that are different from, but entirely in accord, with the partial lawlessness of which Ulanowicz speaks. First, we can make no probability statements about the evolution of the biosphere by Darwinian preadaptations. Consider flipping a fair coin 10,000 times. It will come up heads about 5,000 times with a binomial probability distribution. But note that we could say ahead of time what all the possible outcomes of the 10,000 flips might be: all heads, all tails, and so forth. That is we could prestate the "sample space" of all the possible outcomes, so we could construct a probability measure over this space. But we seem entirely precluded from making any probability statements about Darwinian preadaptations because we cannot prestate the adjacent possible sample space of the biosphere.

Now notice that by the above reasoning we have arrived at nearly the raw chance, the aleatoiric, of which Ulanowicz speaks, but by a different route. The arising of Darwinian preadaptations can be assigned no probability at all. Unlike Ulanowicz, this discussion does not depend upon causal holes in the fabric of space/time and Elsasser's unique combinatorial heterogeneity—whose echo is found in the nonergodic unfolding of the universe above the level of the atom. Conversely, what I have just claimed does not rule out the causal holes in the fabric of space/time of which Ulanowicz speaks.

Next we can ask: do natural laws sufficiently describe the evolution of swim bladders? If by natural law we mean a compact description available, beforehand and afterward, of the regularities of a process, as Murray Gell-Mann argues, then we can have *no sufficient law* for the emergence of swim bladders. We cannot even prestate the possibility of swim bladders, let alone the probability of their emergence, so how can we have a law for their emergence? Note that we have arrived by a different route at Ulanowicz's claim that laws do not sufficiently describe the unfolding evolution of the biosphere.

Whether we take the Ulanowicz view, or that which I have discussed, the results are radical, as Ulanowicz in part discusses. First, the issue of the existence of complex things such as hummingbirds and flowers becomes an issue. Were Weinberg right,

and the laws of physics entailed the evolution of the hummingbird and flowers, which apparently is not the case, then the existence in the universe of hummingbirds and flowers would be
explained by that entailment. But there seems no way that the
laws of physics entail the coming into existence in the nonergodic universe of hummingbirds and the flowers they pollenate
and that feed them nectar. Thus, in the open universe seen via
this discussion or the similar discussion of the third window,
the very existence of flowers and hummingbirds requires an
entirely different account than that which reductionism might
have offered. In its place, Ulanowicz and I both appeal in part
to autocatalytic mutualisms. Thus, the flower and hummingbird exist because when the bird feeds upon nectar, pollen in
the flower rubs onto the beak of the hummingbird, sticks to it,
is transported to the next flower, then rubs off on the stammen
of the next flower, pollenating that second flower. Had all the
pollen fallen off the beak of the hummingbird before it reached
the second flower, pollenization would not have occurred. It is
by this quixotic fact, the stickiness of the beak for pollen, that
flowers and hummingbirds exist in the universe. Of course,
we may add insects as well for they have hairy legs and they
too pollenate flowers. But the main point is that we explain the
physical existence of the flowers and hummingbirds in the universe by this mutualism. The causal arrows do not point downward to particle physics, but upward to the mutualistic system
and natural selection. This is downward causation, as Ulanowicz clearly points out.

Thus, a powerful consequence of the apparent lawlessness of
part of the universe is that we must radically alter our account
of reality. Existence itself of complex organisms in the universe
is not to be explained by a bottom-up approach, but, at least
in part, by the mutualisms of which the author and I speak,
although on different grounds. In fact, the entire biosphere
is broadly mutualistic, food webs and all, given sunlight and

other sources of free energy and a few simple chemicals. More, physics itself is altered, for organisms alter the biosphere, which ultimately alters the planet, hence alters the orbital dynamics of the solar system and galaxy. If so, Weinberg's hope for final theory cannot be a final theory of the evolution of even the physical universe.

A welter of new questions arise. On my account above, how would we prove that no law sufficiently describes Darwinian preadaptions? How would we prove Ulanowicz's and Elsasser's holes in the causal structure of space/time and the raw aleotoric on this line of reasoning? On both our views, we seem driven toward a post-reductionist science, not to replace reductionism, but in unknown ways, to augment or alter it. Thus, how do the mutualisms of which both of us wish to speak, the very conditions of existence of these organisms or their economic and cultural analogues, come into existence? Coordinated behaviors by the mutualistic partners seems required. How does this coordination of properties and activities arise in evolution? Are there principles that enhance the capacity for evolving organisms or economic goods and services, to complement one another? Mutualisms are nonzero sum games. As biological or economic evolution proceeds and species or goods diversity increases, does the creation of new niches arise faster than the creation of new species or goods in the adjacent possible of the biosphere or economy? If so, the growth of the species diversity of the biosphere or goods in the global economy may increase autocatalytically. Diversity drives increased diversity. Does the complexity of features of these new species or goods increase, thereby increasing the ease of evolving ever more positive nonzero sum mutualistic games such that biospheres increase the total diversity of organized processes that can happen as an average trend? Here is my hoped for "fourth law of thermodynamics" for open self-constructing systems such as biospheres.

I have focused in this foreword on some of my own views that lead, like the third window, toward a need for a post-reductionist science. My own view is that neither the third window nor what I have said is remotely sufficient for what we must begin to do. But new issues are raised. As I said at the start of this foreword, *A Third Window* is a bold, radical, and potentially transformative book. I congratulate Robert Ulanowicz on his breadth, wisdom, honesty, and intellectual generosity in laying out his views. This book is the start of its title: *A Third Window*.

Stuart A. Kauffman
October 20, 2008

Preface

If you look at the world through rose-coloured
spectacles, you cannot tell which parts of it really
are rosy and which parts just look rosy.
—Oliver Penrose, "An Asymmetric World"

"Who thinks that what we have heard constitutes a new paradigm?"

The question was put by my friend Henry Rosemont to students in a graduate seminar on the philosophy of science being held at our laboratory. I had just finished describing for them some of the new perspectives that ecosystems science affords on nature. Henry's question aroused mixed feelings in me. Initially, I was irritated, given my aversion to the overuse of Kuhn's word *paradigm*. There followed, however, a tinge of excitement at the possibility that maybe I had not fully appreciated how much the ecological perspective can alter how we see the rest of the world. Perhaps ecosystems science truly offers a new angle on nature (Jørgensen et al. 2007). Hadn't Arne Naess (1988) proposed that "deep ecology" affects one's life and perception of the natural world in a profound and ineffable way? Although I am not adverse to the transcendental, I do nevertheless expect scientists to exhaust every rational approach to phenomena before abandoning them as ineffable.

So Henry's question opened to me the possibility that the ecological narrative truly amounts to a new paradigm. Had I been more honest with myself up to that point, I would have acknowledged that, for decades, I had already been harboring ambitions to describe an alternative approach to reality. I had never felt, as Stuart Kauffman's (1995) title put it, "at home in the universe" as it had been presented to me over the course of my formal training. In fact, I can even point to a definitive encounter that had motivated me to search over the past forty-five years for new foundations upon which to build a rational description of nature.

I was a beginning freshman majoring in Engineering Science at Johns Hopkins. My high-school education had followed an intense and focused technical curriculum, and I was suddenly intoxicated with the possibility of "rounding out" my education. I jumped with both feet into Philosophy 101, a subject that turned out to challenge me in more ways than I ever could have imagined.

The professor in the introductory course was at the time the president of the American Philosophical Society. In addition, he was an excellent lecturer and an intellect of no small renown. He proceeded to "peel the onion" for me—his favorite metaphor for the nature of the human being. A human, he opined, was like an onion. It appears from the outside to have a core at its center. But if one studies the layers of "uniquely human" characteristics, each trait in its turn can be peeled away as superficial. Succeeding layers are removed, until one discovers in the end that there is no center left. This and his numerous other examples of nominalism and materialism stripped me naked of the beliefs I had carried into the classroom. As an 18-year-old, left-brained youngster of recent immigrant stock with almost no formal exposure to the humanities, what rejoinder could I possibly offer?

Defenseless though I was, I nonetheless found it difficult to adopt the metaphysical picture that was being presented to me. In particular, I bridled at the notion of epiphenomenalism—the idea that higher features of the living realm, such as choice and intention, are mere epiphenomena. Like the light flickering on the screen at a movie show, they were accounted to be illusions, devoid of any true agency. I was unable to abandon my belief that they were active agents in changing the natural world. To dismiss them summarily, rather than attempting to weave them into a fuller scientific narrative, smacked to me of intellectual indolence, if not outright dishonesty. I simply could not embrace any metaphysics that precluded the coherent inclusion of all that was actively shaping the world around us.

Dissatisfied with the received wisdom, I found myself dwelling upon those aspects of my curriculum that fit less than neatly into the prevailing worldview. One early fascination was with a chemical engineering course called Properties of Matter. As taught in my department, the course was mostly a survey of statistical mechanics—how thermodynamic variables and physical properties, such as the internal energies, viscosities, and specific heats of systems could be estimated from knowledge about the distributions of their molecular constituents. I marveled at how matters could be horribly messy at lower scales and yet quite well-behaved at higher ones. As I shall highlight early in the first chapter, the nascent field of thermodynamics presented a major challenge to scientific thinking throughout the middle decades of the nineteenth century. I was particularly intrigued by a suggestion on the part of Ilya Prigogine (1945) that arbitrary ensembles of processes somehow take on a configuration that minimizes the overall rate of production of entropy (commonly assumed to be disorder). I wondered whether the individual processes might be responding to some necessity at a larger scale.

It is one thing to accumulate sundry observations, but, failing any core "discovery" around which such fragments could coalesce, my search remained desultory. Fortunately, matters began to focus for me in the late 1970s. My route from chemical engineering into ecology had been one of trying to adapt mechanical models of chemical kinetics so that they might simulate ecosystem behaviors. I even recall one Faustian moment when I stood at the end of our lab's research pier and directed my gaze up the Patuxent estuary, thinking, "If only I could measure all the biological parameters [e.g., copepod feeding rates, sedimentation rates, rates of carbon fixation by algae, etc.], I then could construct a model that would tell me how the estuarine ecosystem would respond to any new combination of conditions." Unfortunately, I was immeasurably far from being able to construct such a model, and my experimentation with much simpler mechanical models had left me quite frustrated and dissatisfied. It hardly seemed I was alone, however, because few elsewhere seemed to be enjoying much success with whole ecosystem models. So, when I was recruited by Trevor Platt of the Bedford Institute to join with similarly disillusioned modelers under the aegis of the Scientific Committee on Oceanic Research (Working Group 59, to be precise), I signed on with enthusiasm.

The members of WG 59 agreed that models built around a single process would often perform satisfactorily enough. A consensus quickly precipitated, however, that mechanical models of several coupled biological processes almost always seemed to go awry (Platt, Mann, and Ulanowicz 1981). Furthermore, the group felt that too much effort had been expended estimating stocks of populations, while not enough attention was being paid to measuring rates of processes and flows. A key recommendation of WG 59 was to shift emphasis in biological oceanography away from describing and estimating (collections of) objects in favor of concentrating on transformations and flows among participating taxa. Through interactions with my colleagues in WG 59, the focus of my own investigations moved away from objects and toward *relationships* among objects.

While doing background reading for the group report, I chanced in close succession upon two seminal papers dealing with the application of information theory to ecological organization (Atlan 1974; Rutledge, Basorre, and Mulholland 1976). By coincidence both papers dealt with what is called the "average mutual information" (AMI) of ecosystem configurations. Essentially, AMI is a measure of how well organized or determinate a configuration of relationships appears, as will be elaborated in the chapters that follow. The mathematical form of the mutual information resembled a familiar quantity from thermodynamics called the Gibbs-Helmholtz free energy, which was constructed to measure how much work a system could possibly perform (Schroeder 2000). The problem was that the AMI, coming as it did from information theory, carried no physical dimensions; it could not indicate the size of the system to which it was being applied. In order to maintain the parallel with thermodynamics, I needed to impart the dimensions of work to the AMI. Perhaps the simplest way of doing this was to scale (multiply) the AMI by the total activity (sum of all flows) inherent in the ecosystem.

The resulting product I called the system's *ascendency* because it represented the coherent power a system could bring to bear in ordering itself and the world around it.[1] Over the course of the following two weeks, I tested how well the measure could mimic various facets of organization. I was excited to discover that the index nicely encapsulated almost all the major attributes that Eugene Odum (1969) had used to characterize more "mature" or developed ecosystems. That is, increasing ascendency appeared to describe quantitatively both the growth and development of ecosystems. As it turned out, I finally had formulated a phenomenological statement around which to configure my accumulated renegade observations.

I soon became aware of my inability to devise any explanation by which ecosystem development in the guise of increasing ascendency could be explained fully in terms of the actions of its individual parts. It gradually dawned upon me that the tenet of increasing ascendency, like the second law before it, directly challenges the prevailing mechanical view of the world. My readings in thermodynamics had alerted me to the fact that, in any confrontation between phenomenology and theory, theory remains at risk, until it can be otherwise supported. Having not yet formulated a coherent theory to elucidate the rise of ascendency, I acted conservatively by presenting my discovery primarily in phenomenological terms. Thus, my first book, *Growth and Development*, carried the subtitle *Ecosystems Phenomenology* (Ulanowicz 1986). In that volume, I also elaborated a number of ancillary mathematical methods useful in analyzing ecosystem networks.

To say that phenomenology is disdained by most biologists is a clear understatement. Furthermore, because *Growth and Development* included considerable algebra, it failed to attract much of a readership among biologists. It became incumbent upon me, therefore, to probe deeper for the causes behind increasing ascendency and to articulate them more in prose rather than in mathematical script. The results of my continuing studies appeared a decade later as *Ecology, The Ascendent Perspective* (henceforth, EAP) (Ulanowicz 1997). The double entendre in the title was intentional and hinted at my growing awareness, fed by Rosemont and shared by Odum (1977), Bateson (1972), and Naess (1988), that ecology had something fundamentally different to tell the world. In that work, I gave heavy emphasis to the nonmechanical attributes of autocatalytic behavior, which I envisioned as the principal source for increasing ascendency. I stopped short, however, of declaring outright that increasing ascendency is at odds with the foundational assumptions of science as we know it.

In the immediate wake of writing EAP, I set about to investigate exactly what assumptions constitute the foundations of science. I discovered axioms that had precipitated during the century following publication of Newton's *Principia*. These were channeled in large measure by the way in which Newton had formulated his laws of mechanics (but not, somewhat ironically, by Newton's personal beliefs). From their apogee early in the nineteenth century, the elements of the Newtonian consensus have all eroded to various degrees due to challenges arising out of thermodynamics, evolutionary theory, relativity, and quantum physics, until only a tattered remnant survives today. Some of the remaining threads are defended vigorously by various biologists (as will be discussed later), but the larger body of scientists remains indifferent to foundations, content simply to regard their erosion as an inevitable casualty of the postmodern, deconstructivist era. That is, few seem to think it possible, or even desirable, to attempt to replace the threadbare Newtonian fabric. Most appear content to let technological progress take its course in abstraction of any underlying metaphysics.

One school that eschews such indifference consists of the postmodern constructivists (Griffin 1996), among which I number myself. Like postmodernists in general, this subgroup affirms the passing of the modern synthesis. The constructivists, however, believe that new foundations can be cobbled together by mixing remnants of the Newtonian era with both the notions of antiquity and radical elements of contemporary thought. My introduction to the school came during a visit to the University of Georgia, where I met one of the prominent exponents of postmodern constructivism, Frederick Ferré. Possibly even more influential was a meeting I had during the same visit with Eugene Odum, the proverbial "grandfather" of American ecology. The day before meeting Odum I had delivered a lecture that outlined the Newtonian assumptions. After breakfast with Odum the next morning, he asked me to list the Newtonian precepts on one side of a piece of paper. After I had done so, he challenged me to fill out the right hand side with how ecosystems ecologists might regard each of the Newtonian tenets. His conviction that ecology causes one to see the world very differently became unmistakably obvious to me.

Gradually, with the help of past thinkers, such as Walter

Elsasser, Robert Rosen, and Gregory Bateson, and through inter-action with a host of friends and colleagues with whom I converse online, I became convinced that the process of ecosystem devel-opment violates *each and every one* of the five postulates upon which the Newtonian worldview rests. I, therefore, formulated and published an "ecological metaphysic" that was cast initially by inverting each of the traditional tenets (Ulanowicz 1999a). Unfortunately, this wholly deconstructive approach violated the spirit of postmodern *constructivism*. To set matters aright, I have struggled over the last several months to exposit in a declarative way those foundations minimally necessary to deduce the sce-nario of increasing ascendency.

I ask the reader to note the sequence of events: I started simply with the desire to examine alternatives to the prevailing metaphysics. Interesting as those exceptions were, they did not of themselves fall into a coherent narrative. It was not until I chanced upon a phenomenological precept (increasing network ascendency) that I discovered a kernel around which a host of ideas (ancient, modern, and contemporary) cohered into the exposition that follows. It is my hope that the ensuing narrative satisfies the ends toward which ecologists, such as Gene Odum, have been striving for most of the past century.

My wife, Marijka, likens the discomfort I experienced as a freshman to a grain of irritating sand that was introduced into my previously comfortable shell. Over the years, I have ever so gradually secreted layer after layer of mother of pearl. As an erstwhile engineer, I think more in terms of networks and com-pare my discovery of increasing ascendency to stumbling upon the outflow of a large river into its estuary (like that outside my office window). Over the years, I have labored upstream through ever-branching tributaries, striving to reach its head-waters. In what follows, however, I will reverse that sequence and begin at the headwaters, traveling with the reader down-stream to trace out the dendritic logic that connects source to outflow. The reader is left to judge whether the subsequent narrative is of whole cloth and whether ecology offers a more coherent and encompassing vision of nature than heretofore has been possible.

The preparation of a book manuscript is a long and tedious process that necessarily diverts one from the crush of more immediate concerns. I therefore wish to thank my two immediate directors of the Chesapeake Biological Laboratory, the late Kenneth Tenore and, now, Margaret Palmer, for being patient with me over the protracted interval during which I gave priority to writing this book over my obligations to help keep our laboratory solvent. Looking back over the years, I also wish to express my gratitude to the late Eugene Cronin and my colleague Joseph Mihursky for trusting that I could manage the transition from engineering into ecology. Appreciation also goes to the late Bengt-Owe Jansson for providing me with several opportunities to promulgate my evolving ideas to important and influential audiences. Written comments on an early draft of this work were gratefully received from Robert Christian, James Coffman, Daniel Fiscus, Sven Erik Jørgensen, Alicia Juarrero, Stuart Kauffman, and an anonymous reviewer solicited by Oxford University Press. Several colleagues provided valuable comments helpful in reworking particular passages. These include Eric Chaisson, Philip Clayton, Sally Goerner, Steven Nachmanovitch, Thomas Robertson, and Philip Welsby. I have endeavored to cite among my references as many of those friends and colleagues as possible that have sustained me through their friendship and discussions. Among those who helped my recent career in peripheral ways but whom I was unable to fit into the thread of my narrative were Luis Abarca-Arenas, Francisco Arreguin-Sanchez, Andrea Belgrano, Antonio Bodini, Ralph and Mary Dwan, Sheila Heymans, Daniel Hoffmann, Thomas Nadeau, Joanna Patricio, James Proctor, Ursula Scharler, Charles Sing, and Frederik Wulff. To all the above, I am deeply grateful for their ideas and their continuing friendship. It is necessary to emphasize that several of the above-named disagree strongly with some of my conclusions. Hence, no one listed should be held accountable for any deficiencies in this work.

It is customary to thank one's spouse at the end of any list of supporters. It would be a grievous understatement, however, for me to cite merely her emotional support. My wife, Marijka, has been a true coworker through the years. She is thoroughly familiar with all the major directions that I am expositing and has at times suggested new approaches. In her critiques, she has constantly urged me to provide illustrative examples to support my points and to compensate for my tendency as an engineer to remain *wortkarg*, or spare of language. Her love, devotion, and indulgence over the many years have been absolutely essential.

Finally, I am writing this valediction with my eye firmly on the future. Our world is experiencing enormous difficulties in coming to terms with its complexity—a circumstance that, I fear, owes in no small measure to the influence of outworn fundamental assumptions. I beg the reader to forgive my hubris in thinking that I can in some way help to correct the disparity between prevailing assumptions and reality. I fully realize the gravity of this endeavor, but I am driven by the hope that what I am proposing might help to reconcile us with the demands progressively being laid upon our evolutionary humanity.

Finally, I am acutely concerned with how our children and their children will be able to cope with the pressures and exigencies that the future brings. It is in the spirit of hope, therefore, that I dedicate this work to my own children, Anastasia, Peter, and Vera, with much love and fatherly devotion.

Port Republic, March 30, 2007

1

Introduction

> "If I am right, the whole of our thinking about what we are and what other people are has got to be restructured. . . . If we continue to operate on the premises that were fashionable in the pre-cybernetic era, . . . we may have twenty or thirty years before the logical reductio ad absurdum of our old positions destroys us."
>
> —Gregory Bateson, *Steps to an Ecology of Mind*

A Self-Destructive Avenue?

The late Gregory Bateson seemed convinced that society is on a suicidal course and that we can be saved only by eschewing our modernist hubris in favor of "an ecology of mind." In effect, Bateson was arguing that the fundamental assumptions that support how we presume the world to function are categorically wrong—not simply askew or in need of amplification or clarification—but outright *wrong!* His assertion surely will strike many readers as preposterous. A look in any direction at any time over the past three centuries reveals major advances and benefits that have accrued to society from adopting the scientific, rationalist perspective. How could such marvels possibly have derived from mistaken foundations? How could continuing to look at the world through the same helpful lens possibly lead us astray? Surely, Bateson was delusional!

But Bateson may seem delusional only because his view of

nature originated from within the scientific community. As C. P. Snow (1963) observed, society is pretty much divided into two cultures with clashing opinions as to whether science affords a beneficial window on reality. Any number of writers, romanticists, and humanists have warned society over the years that the scientific viewpoint illumines only the road to perdition, and, for many, the horrors of the twentieth century proved that point. Goethe (1775) even went as far in *Urfaustus* as to compare placing one's faith in the Newtonian approach with selling one's soul to Evil. More recently, this attitude has drawn succor from postmodern deconstructivists such as Feyerabend (1978). So Bateson has quite a bit of company, it would seem. What distinguished Bateson from most of his fellow critics, however, was that he set out to construct a rational, alternative picture of nature.

That ecology played such a prominent role in Bateson's alternative is highly significant. To be sure, the ever-burgeoning catalog of ecological ills could be taken as part of the very decline that Bateson had prophesied, and he was grieved by these natural maladies. But Bateson made abundantly clear his distance from the attitude that "technological thinking caused the problems; technology can solve them." Such would represent what Bateson called a "pathology of epistemology" (Bateson 1972, 478). Rather, he was calling for a complete overhaul of how we look at the world, one informed by the image of the ecosystem rather than that of a machine. During his lifetime, he made progress toward articulating this new direction by invoking the nascent science of cybernetics and showing how counter-intuitive phenomena could be understood in terms of indirect effects resulting from feedbacks and the connectedness that is characteristic of ecological systems.

Bateson was daring in his suggestion that nature was dualistic, albeit not in the sense of Descartes. Borrowing (perhaps unadvisedly) from Jung's neo-Gnostic vocabulary, Bateson identified as *pleroma* those entities that were homogeneous, continuous and governed by matter and energy—the normal "stuff" of science. Living systems and similar physical analogs that were characterized more by individual differences (information) and reflexive actions he called "creatura." Although he eschewed the transcendental, he nonetheless despaired of how the modern mind-set denies one access to the "sacred" in the natural world around us (Bateson and Bateson 1987). Despite these contributions, it cannot be said that Bateson achieved a full description of what, for want of a better term, might be called an "ecological metaphysic." It is my aim in this book to continue Bateson's agenda and to suggest a complete but rational replacement for those foundations that first initiated and subsequently sustained the scientific revolution. This latest revolution is a call to rational *metanoia*, a thoroughgoing conversion of mind.

Bateson sensed that ecology was not merely a derivative science, one wholly dependent on physics and chemistry for its explanations. Rather, to him ecology afforded a truly different way of perceiving reality. Others have sensed that ecology is fundamentally a different endeavor. Arne Naess (1988), for example, emphasized that ecology was "deep," and he purported that encounters with the ecological affect one's life and perception of the natural world in profound and ineffable ways. Jørgensen et al. (2007) likewise point to a number of attributes of ecosystems that deviate from the conventional and prefigure the discussion that will follow. The complexity of ecological dynamics has prompted some investigators to recognize the necessity for complementary narratives of the same phenomena (Jørgensen 1992). Even outside the discipline, there are those who recognize that ecology offers special insights into other natural and even artificial phenomena: witness, for example, books on the "ecology of computational systems" (Huberman 1988) or the establishment of institutes devoted to the "ecological study of perception and action" (Gibson 1979).

Ecology, the Propitious Theater

What, then, is so special about ecology, and is it indeed as ineffable as Naess would have us believe? I hope I am not spoiling the plot when I state at this early stage that a penetrating read of ecology reveals that it completely inverts the conventional assumptions about how things happen in the natural world. Furthermore, while recognizing the essential mystery surrounding all things living, I would submit that the reasons that ecology is so special are nowise as ineffable as Naess contended. It is possible to identify in perfectly rational fashion where, how, and why ecosystems behaviors depart from conventional dynamics (Ulanowicz 1999a) and to use those essential differences to build a more logical and coherent perspective on the phenomenon of life than can possibly be achieved by looking through the Newtonian glasses.

As the title of this book suggests, I am proposing that, if we are to understand and to survive, it becomes necessary to open a new window upon reality—a third window, so to speak. Without ignoring contributions out of antiquity, one could argue that the first modern window on nature was framed by key figures, such as Hobbes, Bacon, Descartes, and especially Newton, during the era leading up to the Enlightenment. As we shall argue, what one sees out this window was shaped largely by the ideas of Plato and the Eleatic school of fundamental essences. The second window signaled a shift from "law" to "process" and introduced secular history into the scientific narrative. It was opened in two stages, first by Sadi Carnot (1824) and again later by Charles Darwin (1859).

In contrast to these first two windows, the third panorama, that of ecology, has opened more gradually and, some would say, more fitfully. Ecology arose in the latter nineteenth century as certain ideas originating in the then-burgeoning field of physiology were extended beyond the scale of individual organisms. It took a particularly radical (some would say, subversive) turn during the early twentieth century when American Frederic Clements (1916) described ecological communities as akin to organisms. (Clements' detractors are wont to focus upon his chance and somewhat offhanded comment that ecosystems can be regarded as "superorganisms.") Clements' hint that top-down influence might be at work in communities did not at all sit well with conventional thinking, and a contemporary of Clements, one Henry Gleason (1917), countered that ecological ensembles come into being more by chance than by existing regularities. Gleason's view eventually supplanted Clements' during the 1950s, when society focused emphasis upon the action of individuals (Hagen 1992).

Clementsian notions were never entirely eliminated from ecology, however. G. Evelyn Hutchinson (1948), at about the time that cybernetics came into vogue, pointed to circular configurations of causal action as being a key driver behind system-level behavior in ecosystems. In making the case for circular causality, Hutchinson drew upon the work of one of his students, Raymond Lindeman (1942), who gave didactic form to interrelationships within ecosystems by portraying them as networks of transfers of material and energy.[1] Lindeman's graphical approach to ecosystem dynamics was adopted and elaborated by another of Hutchinson's students, Howard T. Odum (1971), who also echoed Hutchinson's opinion that the role played by reward loops in ecosystem development is a highly significant one.

Backing into a New Road

It was precisely this heady mix of whole-system behavior, stochasticity, cybernetics, and networks, as attractively summarized by Eugene and Howard Odum (1959) in their popular textbook *Fundamentals of Ecology*, that first beckoned me and so many other physical scientists to become systems ecologists. To me, ecology seemed such a vibrant and fecund domain in comparison with the nonliving systems that had been my preoccupation as a chemical engineer. Still, it remained a rather inchoate brew to me, and I daresay even to the leading thinkers of the time, as we will see with Bateson. I would like to claim that all these elements fell rapturously into place during one flashing moment of insight, but it did not happen that way. There was a decisive moment, but, rather than being one of insight, it came as a singular juxtaposition of several ideas that led to an exciting *phenomenological* discovery. *Phenomenology*, as used in science, means the encapsulation of regularities into a quantitative formula, achieved *in abstraction of any eliciting causes*. Hence, phenomenology does not imply understanding, although it often leads in that direction.

I had the good fortune to read two papers in close succession (Atlan 1974; Rutledge, Basorre, and Mulholland 1976) that together provided me with a method to quantify the degree of organization inherent in any collection (network) of interacting processes. This discovery itself proved to be highly useful for assessing the status of an ecosystem, but there was more. The mathematics used to quantify organization was borrowed from the discipline of information theory. The measurement of information is accomplished in a strange, converse fashion whereby, in order to assess how much is known about a situation, it is first necessary to quantify its opposite, i.e., how much is unknown. (See chapter 5 in Ulanowicz 1986.) Thus it was that my utilitarian search for a measure of dynamical order brought me into contact with a way of parsing reality that has significant philosophical implications: using information theory, it becomes possible to decompose the complexity of any scenario into two separate terms, one that appraises all that is ordered and coherent about the system and a separate one that encompasses all that is disordered, inefficient, and incoherent within it. Furthermore, the mathematics of the decomposition reveals that these two features are strictly complementary. That is, under most conditions, an increase in either implies a decrease in the other. This agonism revealed for me a fundamental feature of reality that remains absent from virtually all scientific narratives, namely, that nature cannot be regarded in monist fashion. Overwhelmingly, scientists concentrate on elucidating the rules that give rise to order and coherence, but, in complex situations (such as living systems), such explication is never independent of the related dynamics of chance and arbitrary phenomena.

These considerations about the dual nature of reality will be discussed in greater detail toward the end of chapter 4; suffice it for now to remark that they are cogent to Bateson's (1972, 164) dismay over our usual approach to problem solving. Our contemporary predilection is to define a problem, formulate a desired endpoint, and then calculate in monist fashion how most directly to achieve that endpoint. All this is attempted without regard for the dynamics of countervailing aleatoric phenomena, the effects of which propagate over the same network of relationships as do the dynamics that build structure. Everyone is familiar with the unexpected and/or counterintuitive results that can occur when one neglects indirect causal pathways within a complex network of interactions: for example, the DDT used to kill agricultural pests winds up decimating populations of predatory birds. To a degree, such indirect effects can be written into the monist calculus of contemporary problem solving. What is more subtle, however, and absent from the conventional approach is the *necessary* and somewhat paradoxical role that chance and disarray play in the persistence of complex systems, because, without them, a system lacks the flexibility necessary to adapt and becomes defenseless in the face of novel perturbation. This relationship between the complementary dynamics of organization and chance is akin to a Hegelian dialectic. They remain antagonistic within the immediate domain, but they become mutually dependent over the larger realm. Our inclination under the monist approach is to drive the aleatoric to extinction, but to do so beyond a certain point is to *guarantee* disaster.

Turning Around and Going Forward

If we wish to avoid a bad end, then maybe, just maybe, we should pause and reconsider our directions. The foregoing considerations suggest that we may harbor an inadequate or inaccurate image of reality, and so we might begin by scrutinizing our (mostly unspoken) assumptions concerning how nature acts. Although a legion of books is available describing the scientific method, works that elaborate and critique the underlying postulates (metaphysics) of conventional science remain scarce by comparison. This book is an attempt to help redress that imbalance. As the first step toward correcting this bias, I will attempt in the next chapter to delineate the assumptions that frame the two great windows through which we currently regard physical reality—the Newtonian and Darwinian worldviews. With respect to the Darwinian narrative, I will argue in favor of the little heralded shift whereby Darwin's focus on indeterminate "process" effectively replaced the Newtonian concept of "law" as regards living systems. I will argue further that neither window provides an adequate resolution of the complementary (conflicting) questions "How do things change?" and "How do things persist?"

If the conclusion that the conventional windows do not provide an adequate aspect on the world seems too pessimistic to some readers, I would ask them to be patient. This book is not an antiscience screed. In the chapters to follow, I will attempt to construct a rational basis for what I consider to be a more realistic approach to the study of living systems. Should that goal

sound ridiculous and hubristic to some, I would beg them consider the precedent set by Tellegen's theorem in thermodynamics (Mickulecky 1985). Bernard Tellegen worked with network
thermodynamics, where systems of processes are represented
as networks. Each node in the net is characterized by a potential (such as voltage or pressure), while the transfers connecting
the nodes (the arcs or links) are quantified by the magnitude of
the associated flow (amps, m^3/s, respectively). In the conventional view, agency resides in the nodes, and flows are driven
from nodes of higher potential to those with lower values. Thus,
electrical current flows in a radio circuit at the behest of the differences in electrical potential (voltage) between components,
while drinking water flows in a municipal distribution network
in response to differences in hydraulic pressure.

Tellegen discovered that, whenever the relationships between
potentials and flows are strictly linear, system dynamics become
entirely symmetrical as regards the potentials and the flows. That
is, nodes and flows become completely interchangeable; there is
no reason that flows cannot be considered to be the causes of
the given potentials. From this perspective, the convergence of
electrical currents drives up the potential at that intersection
(node). Water pressure may rise at an intersection of lines in a
municipal system because water is arriving there faster than it
is flowing away. In brief, Tellegen showed that throughout the
realm of linear dissipative systems, there are always two identical and inverse (dual) perspectives on the same problem.

Of course, ecology is hardly a linear world, and no one should
expect to achieve a fully equivalent description of ecosystem
behavior by considering flows as causes. This is not as much of
a loss as it first seems, however, because full equivalence would
actually provide no new insights. If, however, a description of a
nonlinear system should become possible whereby flows serve
as causes, it follows that the ensuing picture would differ (possibly markedly so) from the one drawn with the focus on objects.

Furthermore, those differences would not have been visible through the conventional lens. The new perspective affords the opportunity to view situations that are *wholly new*. Such a new vision is exactly what I am trying to convey in this book: an alternative (dual) description of our natural world can indeed be made in terms of processes as causes.

Doubtless, some will object that I get carried away at times with narration in terms of processes at the expense of visions through the more familiar windows. Should the reader be reluctant to make a clean break with historical foundations, I would hope that he or she would at least entertain the feasibility of viewing phenomena through multiple windows in order to obtain a "stereoscopic vision" that might provide new depth to our understanding of nature.

Obviously, we are treading on unfamiliar ground here, and I hope that the reader will accept the narrative that follows in the spirit of "postmodern constructivism" that I mentioned briefly in the preface (Griffin 1996). While the name of this school might sound like an oxymoron to some, it is only because far too much of postmodern critique has consisted solely of deconstruction. However, a relative few among the postmodernists are picking up elements from among the rubble left by deconstructionists and using them to build new ways of visualizing reality. Although narrative no longer requires that one abide by all the Enlightenment restrictions, neither should one forsake rationality in the process. Viewed in a positive light, the postmodern critique frees the investigator to search among classical, Enlightenment, and contemporary thought for concepts that can be woven into a coherent rational whole. Accordingly, in the third chapter, I reach as far back as Aristotle to reconsider the types of causes at work in the world. At the other extreme, I will explore the more recent insights of Walter Elsasser (1981), who argued that our prevailing notions of chance are woefully deficient. His take on radical chance will prompt the first postulate in what will become an alternative "ecological metaphysic." Our awareness of the expanded domain of chance will lead us to question whether physical-like forces or mechanisms play an exclusive role in scientific explanations. One particular alternative is Popper's (1990) broader notion of "propensities," which could provide a more appropriate glue for holding the world together.

I do hope that the new abstractions, such as Popper's propensities, will appeal to the reader, but they probably will not remain in one's lexicon unless someone demonstrates how such entities could possibly originate. This I will attempt to show in chapter 4, and I will also argue there how combinations of propensities can impart form and stability to communities of living beings. Two additional fundamental postulates will be required to support this supposition, and they will complete my triad of primary assumptions. Building upon these axioms, I will endeavor to show how the behaviors of developing systems at times violate each and every postulate that had sustained the Newtonian worldview. Along the way, I also hope to exorcise a neglected mystery that lies at the heart of the neo-Darwinian narrative, namely, from what source does the striving that animates competition among organisms originate (Haught 2003)? Finally, the background provided by this discussion on the maintenance of order will allow me to introduce the fundamental phenomenological concepts of ascendency and overhead that form the kernel around which all notions in this book have been situated. These dual attributes lead naturally into an appreciation of the dialectical or transactional characteristic of nature.

The core of what I call *process ecology* is presented in chapter 5, where I consider the agencies at work in ensemble systems, such as are found in ecology, immunology, sociology, or economics. There I argue that we need to shift emphasis away from objects and focus rather upon configurations of processes. This new perspective on the nature of evolution provides a very different slant from what currently is being promulgated by scientific fundamentalists such as Richard Dawkins or Daniel Dennett. I will return yet again to the complementary relationship between ascendency and overhead to elaborate still further and quantify (to a degree) the "transactional" (dialectical) viewpoint.

By the time we reach chapter 6, we will need to pause and summarize the "ecological metaphysic" that was formulated in rapid succession during the earlier chapters and to compare it point-by-point with the remnants of the Newtonian vision that it replaces. The evolutionary story that process ecology tells will then be seen as an expansion of the Darwinian narrative in ways that Darwin had initiated, but which his successors have largely abandoned. Because this new metaphysic is bound to be controversial, I will attempt to anticipate and counter as many potential criticisms as possible.

In the final chapter, we will peer through the new window framed around process ecology to speculate how the alternative perspective might affect our views on age-old philosophical questions, such as free will, individual responsibility, and the origin of life. We will explore whether process ecology suffices to heal the breach between C. P. Snow's (1963) "two cultures" and whether it circumvents the barriers to the "sacred" in nature that Bateson had so lamented. I will even take the risk of considering whether process ecology might substantially mitigate the sometimes noisy and vehement conflicts between science and religion.[2] We will close with an overview of how process ecology and the ecological metaphysic shift the very groundwork for how we perceive nature and consider whether this dualist vision might not change our attitude toward cosmology from one of unrelenting despair to sound a note of cautious hope.

Now it is time to take stock of the foundations upon which three hundred years of astounding advances in science and technology have rested.

Two Open Windows
on Nature

Your father is the worst kind of pain in the neck—a know-
it-all who's sometimes right. . . . He's a dangerous man!
—Paul Schrader, screenplay for *The Mosquito Coast*

Radical Change Needed?

That an intellect like Gregory Bateson would categorically reject
the premises upon which modern science rests is enough to
strain one's credibility. How could such a gifted thinker entertain such a radical notion? Given his credentials, one cannot
simply dismiss him as a misguided misanthrope or an ignorant
Luddite, but even an outstanding reputation does not make
it any easier to accept the contention that the enterprise that
has relieved so much misery over the past three centuries rests
somehow upon shaky or specious premises. But that indeed is
what Bateson intended, and it forms the crux of this book.

Part of the difficulty some readers may have in coming to
terms with a flawed metaphysic derives from their conviction
that science lays claim to a special type of knowledge—one that
can be verified as no other way of knowing can. And here I wish
to make it perfectly clear that I will not be critiquing scientific
methodology, which, with rare exceptions, remains remark-
ably robust. However, as someone with deep respect for the

phenomenological approach to nature, I note that all the facts about which we remain justifiably confident do not necessarily impart credence to the common assumptions about the fundamental realities of nature. The same facts might rest as well on some other axiomatic platform. For that matter, it is not even self-evident that all *natural* phenomena can be verified by the protocols we have developed. But I would refer the reader to more renowned postmodern critics for thorough discussions of such limitations (e.g., Grenz 1996). Rather, I wish to begin by focusing on the history of the scientific enterprise or, more specifically, on the nature of science as a historical construct, one which under other circumstances could have developed into a very different, but still coherent and efficacious, body of knowledge.

Perhaps none is more mindful of the historical contingency within science than those practitioners who deal routinely in the abstruse and almost otherworldly realm of particle physics. The strange phenomena one encounters there moved one of the most notable physicists of the twentieth century, the late John Wheeler, to describe the evolution of science in analogy to a parlor game (Davies and Brown 1986, 23). As Wheeler depicts it, scientists are like the invitees to a dinner party. Dinner is late, and the hostess bids the company to entertain themselves with a game. They elect to play the game "20 Questions" in which the object is to guess words. That is, one individual is sent out of the room while those remaining choose a particular word. It is explained to the delegated person that, upon returning, he or she will pose a question to each of the group in turn and these questions will be answered with a simple "yes" or "no" until a questioner guesses the word. After the chosen player leaves the room, one of the guests suggests that the group not choose a word. Rather, when the subject returns and poses the first question, the initial respondent is completely free to answer "yes" or "no" on unfettered whim. Similarly, the second person is at liberty to make either reply. The only condition upon the second person is that his or her response may not contradict the first reply. The restriction upon the third respondent is that that individual's reply must not be dissonant with either of the first two answers, and so forth. The game ends when the subject asks, "Is the word xxxxx?" and the only possible response is "yes." The course of the game remains coherent, whim and chance are involved, but the outcome is definitive, albeit indeterminate.

Wheeler was not implying that all of science is as ephemeral as the missing word in his game. He was reflecting rather on experiments in particle physics back when the raster of possible combinations of particle properties was still incomplete. No one had yet observed a particle with such-and-such spin, flavor, and color, and so apparatus is assembled in just such a way so as to facilitate observation of such a particle, if it exists. The object of the search is observed in the apparatus for a couple of nanoseconds. The experiment is declared a success, articles are written, and prizes are awarded. The question that nags at minds like Wheeler's, however, is whether what was observed actually exists in raw nature or whether its existence had been facilitated (constructed?) by the experiment itself.

Qualms like Wheeler's do not seem to predominate among scientists, and many, if not most, believe that scientific research results in hard, objective, if not absolute truth. Such deep confidence was evident when Carl Sagan and others were asked to formulate messages and signs that were to be engraved onto Pioneer I, the first space probe to leave the solar system, in the event the probe were to be discovered by extraterrestrial intelligent beings. The group appropriately commissioned images of naked male and female *Homo sapiens*. They also chose to inscribe the Balmer series of the hydrogen atom in binary characters.[1] The tacit assumption behind the choice was that, regardless of what beings might intercept Pioneer I, if they were intelligent, they would know the universal laws of science, and the significance of the Balmer series would become immediately apparent to them. In other words, the laws and facts of science were considered universal and absolute.

Sagan's conjecture was certainly a rational one, but the possibility remains that whatever "science" the recipients might possess might, in fact, deviate from ours. The sensory organs of the alien beings could differ greatly from our own, as might the history of events that gave rise to their science. Akin to Wheeler's parlor game, the body of science that resulted could be so different from ours that few or no points of contact could be established.[2] Radical incongruity notwithstanding, whatever scientific construct the aliens might possess, it would have to be coherent and efficacious in its own way; otherwise, they would have lacked the wherewithal to retrieve the probe.

Of course, we are not dealing here with extraterrestrial science nor even with elementary particles. My purpose behind these examples is to illustrate the potential influence of historical process upon the interpretation of scientific results. With this point firmly in mind, I now draw the reader's attention to events transpiring during the decades preceding and following what many consider to be *the* key event that ushered in the Enlightenment view of the natural world—Newton's writing of *Principia*.

Historical Preconditioning

The sixteenth century, the "forgotten century" by some historians, was an especially tumultuous and violent time, marked by all manner of bloody strife between parties that mostly were divided by sectarian beliefs. It should not be surprising, then, that maintaining the homogeneity of belief within any particular society became a matter of significant common concern, and such angst afforded special powers to the guardians of the belief structures in the form of what we now would call overweening clericalism. Clerics literally held the power of life and death over their minions, a power that continued into the seventeenth century as well.[3] Because the demarcations between the natural and the supernatural realms had not been firmly established, anyone who made a statement about how nature works potentially exposed him- or herself to intense clerical scrutiny—a condition to be avoided at all costs.

One can discern in the writings of such thinkers as Thomas Hobbes, with his preoccupation about the material constitution of nature, or René Descartes, with his emphasis on the mechanics of reality, a palpable circumspection in how they couched their arguments. It is reasonable to assume that many natural philosophers of the era presented to the public a face that did not entirely mirror their inner beliefs. That such was the case of Isaac Newton is now a matter of record (Westfall 1993). Newton was a confirmed theist, albeit one holding secret heterodox beliefs (Arianism), which, if revealed, could have destroyed him. His habit during the period of his early successes with mathematics and optics had been to weave copious references to religion and alchemy into his narratives. Others, whose personal beliefs may also have deviated from what was sanctioned, included astronomer Edmund Halley and architect Christopher Wren. Fortune was to throw these three talented individuals into a historical encounter (Ulanowicz 1995a, 1997).[4]

In January 1684, Wren and Halley met with renowned mechanic Robert Hooke in Oxford during a session of the Royal Society. Wren and Halley both had been interested in establishing a rigorous connection between the law of inverse-square attraction and the elliptical shape of planetary orbits. When they inquired of Hooke whether such connections were possible, Hooke said that he had described the relationship, but that he intended to keep it secret until others, by failing to solve the problem, had learned to value it (Westfall 1983).

Wren and Halley decided to pursue the matter further, and,

in August of the same year, Halley ventured to Cambridge and sought out Newton to pose the same question to him. Newton told Halley that he too had already solved the problem, but that he had mislaid the proof. Seemingly frustrated, Halley told Wren of Newton's reply, and Wren decided to call the bluff of Hook and Newton (whose mutual animosities were well known) by publicly announcing that an antique book would go as prize to any individual who could provide him with proof of a connection.

Newton apparently did not lie when he claimed that he already had proved the connection, for a copy of such a proof that antedates Halley's visit was found among Newton's papers. Being cautious in the company of someone who communicated with Hook, Newton probably feigned having misplaced the proof. One can only imagine Newton's horror, then, when he looked up his demonstration only to discover that it was flawed. The thought of Hooke's laying claim to Wren's prize apparently drove Newton into a virtual panic. He rushed into seclusion to attempt a rigorous exposition. Once at work, the creative process seems to have overtaken Newton. He virtually disappeared from society until the spring of 1686, when, like Moses coming down off Mount Zion, he emerged on the brink of mental and physical exhaustion with three completed volumes of *Principia* in hand.

In Newton's creative fever to best Hooke, the former lacked time to include his usual links to alchemy and the Divine. He wrote only the barest accounts of the technical elements that comprised his theory. It is not difficult, therefore, to imagine the delight of Edmund Halley (most likely a closet materialist) when he first paged through the draft volumes of *Principia*. In Halley's hands was a first-ever rigorous and minimalist description of the movements of the heavenly spheres without any recourse whatsoever to supernatural agencies. Halley immediately pushed Newton to publish his grand work "as-is." The books immediately caught the attentions of the underground followers of Hobbes and Descartes, and Newton's reputation mushroomed.

Newton inadvertently had become the founding father of a mechanical philosophy of nature to which he himself did not subscribe, for, in Newton's mind, the laws that he was elaborating could in nowise be separated from the Creator and Master of the universe in whom he firmly believed. His efforts at restoring the Divine into his thesis, as appended in the General Scholium that Newton published later, went largely ignored. Reasons that the mechanical worldview precipitated as it did likely are owed to two opposing social trends that suddenly found common cause for making the separation between the mechanical and the supernatural as complete and unambiguous as possible. The first, which we have already mentioned, was the fear of lingering clericalism. Natural philosophers of the time were keen to avoid any hint of the supernatural, lest they touch somehow upon religious belief and thus risk excommunication or extermination. The second thrust was the simple desire to abolish the power of clerics. The latter sought to undermine the authority upon which the power of the clerics rested. It obviously behooved both groups to exposit a radical separation between the natural and the transcendental, so that, in the decades that followed Newton's masterwork, there arose a firm consensus supporting the separation that eventually would bear the name of this genius.

A Precipitating Consensus

According to Depew and Weber (1995), this divorce between the natural and the supernatural took the form of four (tacit and overlapping) points of agreement according to which legitimate science was to be practiced in the wake of Newton. Foremost among them was the assumption of causal *closure*. This postulate enjoined that licit explanations of natural phenomena could refer only to mechanical or material causes. Newton's mechanics quickly became the exemplar as to which causes could legitimately be invoked. Not only is any reference to the supernatural strictly forbidden, but as well is any mention of Aristotelian final cause (see below)—a bane to Francis Bacon. While Depew and Weber did not mention it explicitly, there was a second subassumption that usually escapes notice—namely, that any mechanical cause is necessarily amalgamated with a material one. That is, only those causes that are directly elicited by some form of matter can be deemed legitimate. This corollary assumption means that all causes ultimately reflect the four known forces of physics—strong, weak, electromagnetic, and gravitational—each of which is coupled with a particular form of material.

It is slightly ironic that the first Newtonian postulate be referred to as "closure," because the opening of the first window on reality actually represents the closing or narrowing of attitudes that can be traced back to antiquity. Aristotle's image of causality, for example, was more complicated than the one promulgated by founders of the Enlightenment (Rosen 1985a). Aristotle had distinguished four essential causal forms: (1) material, (2) efficient or mechanical, (3) formal, and (4) final, any or all of which could act either separately or in combination with the others. The textbook example for parsing causality into the four categories is the building of a house. In this activity, the material causes obviously are the stones, mortar, wood, etc., that make up the actual structure, as well as the tools that are used to put these elements together. The efficient causal agents are the workers whose labor brings the material elements together. The formal cause behind the construction of a house is, unfortunately, not as clear-cut as the first two. Its closest analog is the image of the completed house in the mind of the architect, but some point instead to the set of blueprints (Bauplan) that the efficient agents use to organize their construction of the building. The image in the architect's mind follows largely from the needs and purposes of those who are to live in it. Such purposes, for their parts, constitute the final causes behind the building of the structure, e.g., the need for housing by the future occupants or by society in general. The close association of the latter two causes makes it all too easy to conflate formal with final causes in this example.

My colleague Henry Rosemont suggested an alternative example that provides somewhat clearer parsing between the four causes. It is, unfortunately, a rather unsavory one—a military battle (Ulanowicz 1995a, 1997). The material causes of a battle are the weapons and ordnance that individual soldiers use against their enemies. Those soldiers, in turn, become the efficient causes because it is they who actually swing the sword or pull the trigger to inflict unspeakable harm upon each other. In the end, the armies were set against each other for reasons that were economic, social, and/or political in nature—reasons that provide the final cause or ultimate context under which the battle is waged. As for formal causes, they become the juxtaposition of the two armies vis-a-vis each other in the physical context of the landscape and the attendant meteorological conditions. These latter forms are the concern of the officers, whose commands shape the battle. This example has a secondary advantage in that it provides a didactic correlation between the physical manifestations of three types of agency (efficient, formal, and final) and the hierarchical status of those whose respective concern they are (private, officer, and head of state).

Missing from the hierarchical assignment of agencies is material cause, which for Aristotle played a distinctly passive role. That is, material causes are necessary for events to proceed, and their exact natures can affect the outcome of whatever *process* is transpiring. By and large, however, material elements are manipulated by the efficient causes, which remain distinct from them. That is, in Aristotle's view, material and efficient causes remain either independent or loosely coupled. His perspective contrasts markedly with the modern scenario, which makes the amalgamation of material with mechanical causes obligatory.

A second postulate that gained favor in the wake of Newton is related to closure and is the idea that nature is *atomistic*—a concept that can be traced back at least to Democritus (fourth century BCE). Not only did this assumption entail the belief that there exist fundamental, unchanging, smallest material units, but also that these units could be built up and taken apart again.[5] There are at least two thrusts implicit in this tenet. The first, the combination of atomism and closure, gives rise to the notion of reductionism, which holds that the only proper direction of scientific exploration is analytical (i.e., explanations of larger phenomena are to be sought exclusively among events at smaller scales). Thus, Carl Sagan, in summarizing his television show on biological evolution, after having presented captivating images of dinosaurs doing ferocious battle with each other and otherwise interacting, declared, "These are some of the things that *molecules* do!" The second impulse concerns decomposability—that in breaking a system into subunits, nothing of essence is thereby lost. When atomism is combined with closure, the outcome is akin to the dictum of Lucretius (first century BCE): "There are atoms, and there is the void"—nothing more. In spite of the fact that we now know much more about the fundamental nature of matter, many continue to think in terms of the Bohr model of the atom or even in the Lucretian image of "billiard balls and vacuum" as constituting the basic substrate of the world.

That systems are decomposable relates in its turn to yet a third important Newtonian assumption, namely, that processes are inherently *reversible*. While this premise might seem strange to anyone familiar with the impermanence of biological phenomena, it should be noted that, until quite recently, all known physical laws and the equations that describe them were perfectly symmetrical with respect to time. That is, if one were to take a picture of some phenomenon acting under the aegis of a law, it would look the same whether run forward or backward, e.g., the collision of two billiard balls.[6] At the turn of the last century, Aemalie Noether (1983) not only emphasized the ubiquity of this assumption but also demonstrated how such symmetry is tantamount to conservation over time. Since the conservation of one attribute or another (e.g., mass, energy, etc.) is usually postulated in most contemporary scientific investigations, the assumption of reversibility is thereby implicitly invoked. It should be noted in passing that any narrative predicated solely on the strongly conservative formal laws of physics is incapable of addressing true change. In such a neoplatonic, reversible world, nothing essentially new can possibly arise.

In the decades following Newton's astounding success at predicting the movements of the spheres, a number of similar accomplishments in fields such as chemistry and electricity ensued, giving rise to the fourth consensus that the laws of nature, once fully elaborated, were sufficient to predict all phenomena. This means that nature is *deterministic*—given precise initial conditions, the future (and past) states of a system can be specified with arbitrary precision. So enamored of their own successes were the mechanists that, by the early nineteenth century, Pierre Laplace (1814) dared to exult in the unlimited horizons of the emerging mechanical worldview. Any "demon" or angel, he rhapsodized, that had a knowledge of the positions and momenta of all particles in the universe at a single instant could invoke Newtonian-like dynamics to predict all future events and/or hindcast all of history. For Laplace, the apotheosis of scientific power lay in billiard balls and vacuum. There was significant philosophical fallout from the belief in mechanical determinism. For example, if natural laws were sufficient to explain all events, then the notion of humans expressing any sort of "will" becomes absurd.

Finally, yet a fifth postulate was related to the confidence placed in the determinism of the laws of nature, namely, the belief that they also are *universal* (Ulanowicz 1999a). That is, they are assumed to apply everywhere, at all times, and over all scales. Since mathematical physics on numerous occasions has been able to predict phenomena before they were observed, or even imagined, some physicists have taken this to mean that the laws of physics are applicable without regard to scale. Thus, one encounters physicists who talk about "point-sized black holes"—collections of so much mass in infinitesimally small spaces that the gravity the mass generates will not allow light or other radiation to escape. These theoreticians have extrapolated the equations relating mass, gravity, and light arbitrarily close into a mathematical "singular point" (a point where the mathematical behavior becomes pathological), and they believe that the relationships retain their validity in this uncharted realm. A similar extrapolation is the attempt by other physicists to marry quantum theory with gravity, when the characteristic dimensions of the two phenomena differ by some forty-two orders of magnitude (Hawking 1988). Reinforcing the point made earlier, the conjunction of closure with universality implies that nothing truly new can occur. As Hawking put it, there is simply nothing left for any "creator" to do (ibid.).

Challenges Arising

The heady vision of the world at the beginning of the nineteenth century was such that Laplace and colleagues could celebrate the triumph and hegemony of the Newtonian worldview. This honeymoon was not to last for long, however. Although most today associate the demise of the Newtonian worldview with the appearance of relativity and quantum theories early in the twentieth century, it was but a mere six years after Laplace's pronouncement that French engineer Sadi Carnot's report on the behavior of steam in engines threw the dictum of reversibility (and conservation) into serious question (see "Synthesis or Sublimation" below). To this day, Carnot's challenge remains largely unanswered (O. Penrose 2005). We shall see how, later in the nineteenth century, Darwin was influenced by earlier discoveries in geology and was among the first to introduce history and true change into biological dynamics (see "A New Window Opens" below). Then, as most are aware, the development of relativity and quantum theories early in the twentieth century finally enunciated clear boundaries on universality and determinism.

As a result of these challenges, no one talks anymore about inhabiting a Newtonian universe. I freely admit that, by articulating these Newtonian precepts, I intentionally have set up a straw man. I certainly do not wish to insult any reader's intelligence by implying that he or she fully accepts this two-hundred-year-old consensus. To the best of my experience, no one today holds firmly to all five postulates. It is no exaggeration to say that the Newtonian worldview is in tatters. Unfortunately, surprisingly few of us seem willing to admit this condition. It is poignant to ask, therefore, what has arisen that can take the place of the Newtonian framework. As we shall see, there have been a number of thinkers who have suggested fertile new directions, but none has been accorded widespread attention. Rather, what one encounters among the scientific community is that most of us by and large cling to some dangling threads of the Newtonian worldview. It's just that there remains no widespread consensus about how much weight, if any, should be given to each assumption.

Presently, I will commence deconstructing my straw man. I hope thereby to cast strong doubts upon *all five* of the Newtonian articles, so that, no matter at which thread(s) a particular reader might be grasping, he or she will be moved to question the utility of hanging onto them, as opposed to entertaining putatively more reliable handholds that I will soon suggest.

For I worry that most of us have become reluctant to discuss deep assumptions about nature, preferring instead the refuge of a determined pragmatism or technocracy. It's as if fundamental principles are now somehow immaterial to our quest for a more comfortable, healthier life.

Revanchism?

Pragmatism notwithstanding, what to me seems passing strange is that some individuals in fields where the mechanical worldview would seem *least* applicable remain the most ardent champions of an outworn metaphysic. They seem intent on turning back the clock. Thus, we shall soon see how closure is strictly enforced in the neo-Darwinian scenario of evolution. Dawkins (1976) and Dennett (1995), for example, both prominent advocates of this formulation, are scrupulous in making reference to only material and mechanical causes. Atomism (reductionism) continues to dominate biology—witness the prevalence of molecular biology today. A surprising fraction of scientists today continues to deny the reality of chance in the world, contending instead that probability simply papers over an underlying determinacy (Bohm 1989, Patten 1999).

One is moved to ask what inspires such fervid loyalty to the traditional postulates. With some, it likely is the fear of giving up any possibility for ultimate control over their lives and environment. As we shall discuss more fully later, the metaphysic that was crafted to separate the natural from the transcendental has become for some a set of materialistic *beliefs* in their own right and a rather effective weapon that can be wielded against transcendentalists and theists (Susskind 2005)—an instrument not to be abandoned lightly. Whatever the motivations, matters hardly ever play out as one might hope, as we saw with Newton, the transcendentalist whose work became the cornerstone of a hard materialism. The unexpected seems always to raise its head, although sometimes the very nature of a revolution is not appreciated until long after the fact. Thus it transpired that the second major revolution (our second window) was opened by someone with impeccable Newtonian intentions—Charles A. Darwin.

A New Window Opens

Darwin regarded himself as a staunch Newtonian (Depew and Weber 1995). His professed ambition was to become "the Newton of a blade of grass."[7] He was scrupulous in how he externalized the efficient agencies of evolution (natural selection) away from the object on which it acted (the organism). But, as was the case with Newton, desire does not always lead to one's intended goal. Possibly even more than Carnot before him, Darwin succeeded in bumping the Newtonian juggernaut off its tracks.

Darwin's basic scheme was simple enough—so extremely simple that some say the resultant scenario became "colonial" in the sense that it could be applied to almost any situation, even many where it does not apply (Salthe 1989). The basic unit of selection is the individual organism, which somehow is endowed with being able to reproduce itself and to strive against others that might compete with it for finite resources (Haught 2003). During the course of its lifetime, the organism is impacted by sundry external agencies, all of which are lumped under the rubric of "natural selection." Those organisms that survived the rigors of natural selection until the time of reproduction got to pass their characteristics on to the next generation, members of which would repeat the scenario. Each offspring, while it closely resembled its parent(s), might differ from them in slight ways. Those differences that resulted in a greater likelihood of surviving until reproduction would thereby constitute greater proportions of succeeding generations. The entire scenario became known as "descent with modification."

As an erstwhile engineer, I immediately confess to a lack

of schooling in the various arguments that ensued in the wake of Darwin's thesis. I am thus ill equipped to discuss with any erudition such issues as panadaptationism or genetic fixation and surrender those topics to the care and safekeeping of evolutionary biologists. I wish instead to take advantage of my perspective as an outsider to focus on several advantages and deficiencies of the Darwinian scenario that, to the best of my knowledge, have not played prominent roles in most discussions about evolution. For me, these secondary issues help to demonstrate the true depth of Darwin's revolution and at the same time to point to the need for opening yet a third window.

Although Buffon (1778), Cuvier (1825), and Lyell (1830) all had earlier suggested that the past plays a role in the natural world, it is Darwin's reference to history that remains most prominent in our minds today. In reproducing both form and function, members of a species pass on to the next generation the cumulative traits they acquired from previous generations as well as aleatoric modifications that may have occurred at reproduction, i.e., their histories. The particular evolutionary pathway followed by the predecessors of a given organism strongly constrains what will be possible in the next generation. Such temporal constraint differs markedly from the sequences of events created by Newtonian laws, which, as we have remarked, can be read either forward or backward.

Law or Process?

While some do give voice to the role of history in Darwinian evolution, far fewer seem motivated to emphasize that Darwin's narrative represented a radical departure from the absolute lawfulness of the Newtonian world. One doesn't speak, for example, of Darwin's *law* of evolution. Rather, one refers to it as a theory, or less frequently as the *process* of evolution. This shift from law to process is highly significant because, as Whitehead and Russell (1913) were able to demonstrate, the outcome of a proper law is always determinate. With a process, however, despite the action of particular constraints, the outcome is decidedly indeterminate.

Because the notion of process will play a substantial role in the remainder of this book, it is helpful at this point to be as precise as possible about what we mean by the term. Accordingly, we adopt an operational definition:

> A process is the interaction of random events upon a configuration of constraints that results in a nonrandom but indeterminate outcome.

Actually, we have already encountered the action of process in Wheeler's parlor game. Another simple, artificial, but very illustrative example is Polya's urn (Cohen 1976), named after the mathematician György Pólya, who first described it. To carry out the physical process requires an "urn" or opaque container with a mouth narrow enough not to allow one to peek readily inside. One also needs an ample supply of red balls and blue balls that are small enough to enable many of them to fit into the urn. At the beginning of the process, a single red ball and a single blue ball are already in the urn. The urn is shaken, and a blind draw is made from it. If the ball drawn is red, it is returned to the urn along with another red ball, and the contents are shaken again. If the ball drawn at any time is blue, both it and an additional blue ball are placed into the urn. The procedure is repeated for many draws and replacements.

The Polya scenario is very easy to program and simulate.[8] The first question that arises is whether the process "converges," i.e., whether the ratio of red balls to blue balls in the container approaches a constant. Is the outcome nonrandom? It is rather easy to demonstrate that it is. Figure 2.1, for example, depicts the ratio of red balls to blue balls after each of the first one hundred draws. After a full one thousand draws, the ratio of red to blue balls converges to a limit near 0.54591. During the course of any series of draws, the ratio becomes progressively less random.

That this limit differs from 0.5 raises a second question. Is the limit to this process unique, i.e., is it determinate? If the urn were emptied and the process begun anew, would the ratio converge to the same limit? Figure 2.2 shows the first one hundred draws of a second simulation of the process. It is quite obvious that the second series does not converge to the same limit. After one thousand draws, the second series approaches a limit in the vicinity of 0.19561. The Polya process is *indeterminate*.

In fact, if the process is repeated a large number of times, it eventually becomes apparent that the ratio of red to blue balls can converge to any real number between zero and one. A few of the trials converge toward the extremes of homogeneity, as, for example, the one in Figure 2.3. After one thousand draws, it converges to a limit near 0.96806. It becomes clear after watching a

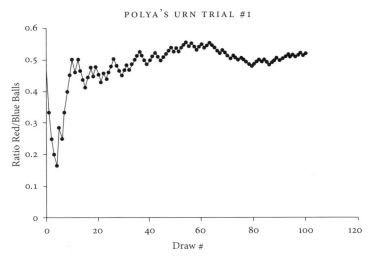

POLYA'S URN TRIAL #1

Figure 2.1. The ratio of red to blue balls after each of the first one hundred draws of a simulation of the Polya process.

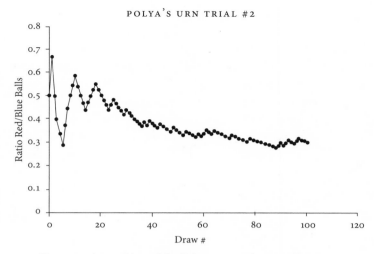

Figure 2.2. A repetition of the Polya process shown in Figure 2.1.

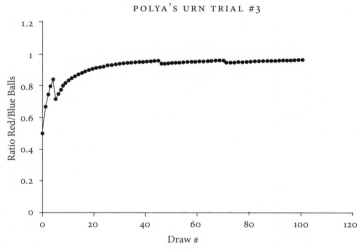

Figure 2.3. A series of draws from Polya's urn that converges
to a mixture heavily dominated by red balls.

number of repetitions that what happens during the first twenty or so draws strongly influences where in the interval 0 to 1 that particular series will converge. In other words, any particular series becomes progressively more constrained by what has gone before, i.e., by its history.

Serendipitously, this simple example teaches us three important things about processes in general:

1. The process is subject to stochastic inputs, but the outcome over time becomes less random (by definition).

2. The process is self-referential in that the ratio of balls at the time of the draw affects the outcome of that draw.

3. The change in ratio of the balls is progressively constrained by the actual history of draws that have already occurred.

These considerations on the notion of process show us that Darwin, despite his affection for Newtonian protocols, set us on a wholly new approach to the living world. He essentially was telling us that living systems arise *not* out of the set of immutable laws that regulate all physical order, but rather as the result of natural *processes* that create living order out of abundant chaos. The late student of mass media Marshall McLuhan (1964) liked to emphasize the mesmerizing effect that familiar ideas and habits have in preventing us from recognizing the truly novel. Thus it is that a significant fraction of ecologists, perhaps out of a misguided sense of "physics envy" (Cohen 1976), continues to search for biological "laws" to illumine the living realm a full century and a half after their venerated intellectual progenitor scoped out for them an entirely new direction in which to pursue the description of life.

The Forgotten Darwin?

As often happens throughout history, the successors of a visionary are usually the ones who edit and interpret what the revolutionary accomplished, and their motivations often differ from those of their mentor. We saw this in the case of Newton, when his works were interpreted in ways that he never intended. Matters transpired somewhat differently in the wake of Darwin. His followers did seek to interpret his theory in the Newtonian tradition that Darwin so highly esteemed. But, by trying to adhere to Newton, Darwin and his followers minimized the truly revolutionary meaning of his work—that *process is more important than law* in shaping living systems. On the other hand, other directions that Darwin suggested were quietly allowed to atrophy because they did not conform to contemporary orthodoxy. For example, he was decidedly sympathetic to the existence of directionality in evolution, and even to outright teleology,[9] yet such empathy is quietly glossed over today. Furthermore, in chapter 4 of *The Origin of Species*, the author laid special emphasis upon "the division of labour" in biological systems,[10] yet this direction/culmination, too, has been virtually forgotten. The division of labor, however, is an apt description of the progressive specialization that we shall be discussing in the chapters that follow.

Not only have various aspects of evolution received emphasis different from what Darwin had intended, but the popularity of his overall scheme has vacillated over the years as well. It may surprise some readers to learn that the theory of evolution, after a stormy entry onto the intellectual stage, waned significantly toward the end of the nineteenth century (Depew and Weber 1995). Gregor Mendel's observations on the discrete characteristics of peas caused many to question the gradual and continuous evolutionary process that Darwin had suggested. Furthermore, Mendel's focus on the discreteness that accompanies genetic inheritance implied that chance was at work in evolution, and new stochastic elements have never been eagerly embraced by the scientific community. So, the spotlight turned instead upon the more predictable among biological phenomena such as comprise ontogeny or developmental biology, which decidedly eclipsed evolution over the next thirty or so years.

Synthesis or Sublimation?

This crook in the road for evolutionary theory is reminiscent of a previous setback encountered by Newtonianism almost a century earlier. As mentioned, the first true challenge to the Newtonian worldview came from engineer Sadi Carnot, who discovered the irreversible nature of heat transfer. His finding eventually became known as the second law of thermodynamics (even though it was discovered *before* the first law [Ulanowicz 1986, 17]). Carnot's principle placed the Newtonian and atomistic theories acutely into jeopardy: if Newtonian-like reversible laws apply to the microscale encounters between the atoms of a gas, how could an ensemble of reversible interactions possibly give rise to irreversible behavior at the macroscale, as the postulate of atomism demands? It was an enigma that occupied many of the best minds in physics for the next half-century.

The "resolution" of this paradox came later in the nineteenth century through Ludwig von Boltzmann, James Clerk Maxwell, and Josiah Willard Gibbs. The three, working independently, considered the atoms of an ideal gas as a statistical ensemble. An ideal gas is one that consists of point-sized particles of mass that act independently of each other. Under the assumption that the momenta of the gas atoms are distributed in normal statistical fashion, Boltzmann was able to show that a particular function of the distribution (a surrogate for its entropy) would rise in a monotonic (irreversible) fashion. Gibbs and Maxwell elaborated a formal theory for such statistical considerations. In what should embarrass any right-thinking positivist, the consensus then arose that this reconciliation, which pertained only to an extremely narrow set of circumstances, applied to the entire universe. Instead of trying to falsify the validity of the Newtonian perspective, as positivism would bid one do, there was a rush to the judgment that no contradiction exists anywhere between the second law and Newtonian mechanics.

In effect, physicists considered that Boltzmann, Maxwell, and Gibbs had "sanitized" the messy phenomenological world of thermodynamics, so that, for physicists, statistical mechanics *became* thermodynamics (in contempt for the phenomenological construct that remains the crux of thermodynamics for engineers).

This redemption of Newtonian thought found repetition in the resurrection of Darwin's theory decades later. As just mentioned, Mendel's discovery of discreteness and chance in heredity raised the specter of irreversibility in evolutionary theory.[11] With chance now rampant on the scene, who would put the genie back into its bottle? The task fell to mathematician Ronald A. Fisher, who was captivated by the elegant statistics that Maxwell and Boltzmann had used in creating statistical mechanics (Ulanowicz 1997). Fisher recapitulated virtually the same mathematics in his demonstration of how the gradualist views of Darwin could be made compatible with the discrete characteristics featured in Mendelian genetics. To accomplish this, Fisher had to assume a large population of randomly breeding individuals (in distinct analogy to the ideal gas assumption). He assumed that characteristics of the organism or phenotype were determined not by a single gene, but by many genes that are changing constantly in Mendelian fashion. Essentially, Fisher tracked the trajectories of gene frequencies in the same probabilistic spirit that Maxwell, Boltzmann, and Gibbs tracked arrays of gas molecules (Depew and Weber 1995).

The culmination of Fisher's labors was his "Fundamental Theorem of Natural Selection," whereby the fitness (reproduction) of gene frequencies is said to be maximized, just as entropy was maximized in Boltzmann's scenario. Fisher noted one very important difference, however. Whereas Boltzmann's system exhibited increasing disorder or looseness, Fisher's gene profiles develop in entirely the opposite direction—that of ever-increasing constraint. This observation moved Fisher to declare astutely that we inhabit a "two-tendency universe," where thermodynamics and genetics evolve in opposite directions (cf. Hodge 1992).

If indeterminacy had been an irritant thrust into the comfortable shell of biology, Fisher had seen to make a pearl of the offending element. Curiously, the situation Fisher had described was the inverse of that in statistical mechanics, where lawful regularity prevails at the microlevel and irreversibility reigns at larger scales. For him, the action of chance was endemic at the microscale of the genome, but selection continued its work at the macroscopic level with the same Newtonian regularity, linearity, and gradualism that Darwin had assumed. In formulating his model, Fisher had clearly demarcated the confines of chance for biology—it was relegated to the netherworld of molecular activities. He also had rescued some hope for a modicum of prediction in the statistical sense of the word. Nevertheless, just as we questioned the reconciliation purportedly effected by statistical mechanics, we justifiably might question whether the eager cohort of biologists that rushed to extrapolate Fisher's "grand synthesis" well beyond the very narrow confines of its assumptions may have acted with undue haste.

A Window Closing Again?

Momentum in Darwin's favor grew in the wake of Fisher's synthesis. Just two decades later, and soon after Schroedinger (1944) had speculated on the molecular character of the genome, Watson and Crick (1953) made their stunning description of the structure of the DNA molecule and how information could be encoded within it. Now, with the "billiard balls" of genomic information clearly defined, the synthesis began to solidify. Descriptions of the mechanisms of meiosis, mitosis, and a myriad of other molecular activities soon followed. Focus in this "neo-Darwinian" synthesis remained decidedly upon the microscale, and the elucidation of the molecular "machinery" became the order of the day. Beyond this domain lay only "vacuum."

This rush of discoveries instigated a revanche of the mechanical perspective. The allure of the mechanistic narrative only grew with the contemporaneous development of computational machinery, which provided a new metaphor for mechanical action—the algorithm. All of biology suddenly appeared to be scripted, and increasing effort was given toward the discovery of constituent rules or mechanisms at the expense of interest in the overall system. Still today, the algorithm is paramount in Daniel Dennett's (1995) view of reality, and he is hardly alone. There is a land-office business in the development of "artificial life," by which is meant the simulation of lifelike behaviors *in silico*. Computers have replaced clocks as the archetype of mechanism, and futurists such as Edward Fredkin (Wright 1988) and Stephen Wolfram (2002) have gone so far as to suggest that cosmology might well be rescripted as the workings of one grand computational machine.

Amid all these stunning new advances, it seems a bit indelicate to ask whether everyone might have been mesmerized by the trees while losing sight of the forest. Might our focus have wandered from the guiding *process* that Darwin had described? Or more precisely, have we become so enraptured by microscale mechanisms that we have lost sight of how Darwin actually described a macroscopic process and not a law? Elsewhere I have discussed the "schizoid" nature of neo-Darwinism (Ulanowicz 1986, 1997)—one's attention is constantly shifting back and forth from the netherworld of the molecule to the macroworld of the organism, as if several orders of magnitude in which a myriad of processes and structures occur matter not one whit. Contemporary focus is upon the *correlation* between specific genomes and large-scale forms and behaviors, as if all efficient agencies resided in the molecular realm. Again, the not-so-tacit assumption here is that of causal reductionism—that all cause must be referred down the hierarchy of scale, or, to repeat Carl Sagan, "These are some of the things that molecules do!"

Obviously, working with genetic codes offers some outstanding technological possibilities, and my purpose is not to gainsay any potential contributions by reductionism in general. Research into genetics will continue to yield significant rewards, as will reductionistic biology. But I pointedly wish to ask whether reductionism is truly all there is to science. Are we indeed satisfied with knowing so little about what transpires across the numerous levels separating the molecular genome from the organism? What if we have misappropriated efficient agency exclusively to the molecular realm, when, in fact, it is being exerted at intermediate levels? Playing with molecules and ignoring dynamics at the mesoscale could lead to gross misunderstandings, if not outright disaster.

The Neglected Middle

And so we turn our attention to that enormous, neglected middle (Liljenstroem and Svedin 2005). Just one step up from DNA, the genomic sequence is read by a network of enzymatic/catalytic reactions (Strohman 2001). Are these networks indeed strictly machines? Are they devoid of agency? Are we really saying that it is the columbic field surrounding the DNA strand that directs all above it? What if the selection of the parts of the genomic code to be read is generated by agency at the next higher level? Recalling Aristotle's causal typology, it is more likely for causality to be distributed across levels rather than be restricted to a single one (the bottommost). Furthermore, we have noted that, in Aristotle's framework, material cause plays a more-or-less passive, secondary role. If his broader, overarching perspective is correct, it would then be a grave mistake to attribute all control to passive agencies while neglecting those that truly direct matters. We need to consider the possibility (likelihood?) that genetic molecules provide merely the material cause, while effective agency is being exerted by the larger networks.

And so the questions continue—all the way to the very top. For example, it is taken for granted in the Darwinian narrative (and I have never seen it emphasized) that over the course of affairs between inception and reproduction, organisms strive to compete for resources (Haught 2003). But what is the cause of this striving? To say that the drive is encoded in the organism's genes explains nothing. The *reductio ad absurdum* of such a suggestion is revealed in a cartoon from the series *Kudzu* by Doug Marlette (Jan. 23, 1997). In it the main character, Kudzu, sits next to his parson and asks, "When will we discover the gene that makes us believe that everything is determined by genes?" Of course, many will retort that it is beyond the ken of science to ask the question "Why?" but I hold that such an attitude sells the scientific enterprise needlessly short. In chapter 4, we shall return to the question of striving and propose a mesoscale explanation.

Finally, it is widely acknowledged and much discussed how the Darwinian process is inadequate to the task of explaining how life originated or, for that matter, how new species come into existence. These lacunae reveal the inadequacy of either of the first two windows to deal with real change. The Newtonian window looks out on an eternally changeless universe. The Darwinian window does reveal a degree of change but falls short of depicting the more radical changes that have structured our universe. To put a finer point on it, Darwinian scenarios do not uncover how true novelty emerges. In order to get the full picture, it becomes necessary to open a still larger window on the world, one that properly frames both the radical indeterminacy and the genuine novelty that occur in the real world.

3

How Can Things Truly Change?

[T]his hope, which has accompanied all the subsequent history
of the world and mankind, has emphasized right from the outset
that this history is branded with the mark of radical uncertainty.
—Ilya Prigogine and Isabelle Stengers, *Order out of Chaos*

Change versus Stasis

If we are to entertain any hope of understanding how things
change in the world (beyond mere change of position), it quickly
becomes apparent that we need to move beyond the limitations
of the Newtonian worldview. Darwin did so, although he and
many of his followers preferred to interpret his theory in retro-
spective Newtonian terms. I mentioned in passing how the classical Newtonian worldview owes much to the Platonic or Eleatic
school of Greek thought that centered discourse on unchanging "essences" as the elements of primary import. But the Eleatic
school did not comprise all of Hellenistic thinking. Opposing
them in a continuing dialectic stood the Milesian school, the
central figure of which was Heraclitus. His famous quotation
"παντα ρει" (everything changes) infers that the only constant in
the world is change. That one can never step twice into exactly
the same river is a common exemplar of this dictum.

As an aside, one sees this dialogue reflected in the heavy influence that the Eleatic position holds over the science of thermodynamics. The thermodynamic variables of primary importance are called "state variables" because they quantify and convey the status of a system when at equilibrium (stasis). State variables have several convenient mathematical properties; primary among them is that they are perfect differentials (meaning that the rules of calculus can be invoked to elucidate manifold relationships among the various state variables). Decidedly secondary status is begrudgingly accorded to "process variables," such as the flow of heat or electricity, because, heretofore, fewer mathematical tools have been available that can treat them. Calculating differences in process variables generated by a changing system becomes problematic because, unlike with state variables, the magnitude of change varies according to the exact pathway that the system traverses. Such "pathway dependence" is a simple reflection of the importance of the system's history. With the burgeoning interest in networks, wherein flows are accorded parity with states (nodes), it becomes likely that the groundwork in thermodynamics may soon shift in favor of flow variables.

It needs to be acknowledged at the outset that full reconciliation between stasis and change is impossible, notwithstanding the fact that *both* are readily observable aspects of nature. (The physicist or engineer might remark that the two concepts are fundamentally different entities, like oranges and apples, because the dimensions of change include time, which is absent from the dimensions of stasis). When looking at nature, one never perceives either pure stasis or complete change. Given sufficient time, no natural object persists indefinitely. Conversely, what superficially appears as change is sometimes conserved when regarded through a different lens (e.g., the momentum of an object moving in the absence of any force).

Against this background, it could be argued that Darwin's

process of "descent with modification" has helped to break the stranglehold of essentialism on scientific thought. The Darwinian window opened science not simply to history but also to process and chance. Change became possible, but, as we saw in the previous chapter, its radius was circumscribed. Darwinian change acts only within type (species), and the process is not open to the generation of new types (speciation) (Gould 2002). Nor has the Darwinian tradition generated any way to "bootstrap" the process of life, i.e., the origin of life remains wholly outside the perimeter of Darwinian explanation.

Chance, the Unwelcome Intruder

The limits on Darwinian theory are related to the antagonism between chance and the goals of science. Whereas science aims to codify, simplify, and predict, the interjection of chance into the narrative results in conspicuous exceptions to regularity, complications in specifying the system, and degradation of the ability to predict. In the last chapter, we discussed two instances (statistical mechanics and the grand synthesis) of how science has attempted to mitigate the challenges posed by stochastic interference. Both reconciliations rested upon the same mathematical tool—probability theory—to retrieve some degree of regularity and predictability over the long run. Probability theory, however, comes with its own vulnerabilities. To apply it forces one to accept a set of assumptions regarding how chance is distributed, e.g., normally, exponentially via power-law, etc.

More importantly still, probability theory can be used only after a more fundamental set of assumptions has been accepted. These essential preconditions are rarely mentioned in introductions to probability—namely, probability applies only to chance events that are simple, generic, and repeatable. Simple events are *atomic* in that they occur at the smallest scale and over the shortest time interval under consideration. *Generic* means that there is no particular characteristic of the events worthy of mention. They are all homogeneous and indistinguishable and lack directionality, save perhaps occasionally exhibiting a binary difference in sign. Finally, it is necessary for one to observe many repetitions of the same chance event; otherwise, one cannot gauge how frequently it occurs (i.e., estimate its probability).

As the cliché goes, if one possesses only a hammer, everything begins to look like a nail! Because probability theory works only on simple, generic, and repeatable chance, most tacitly assume that all instances of chance share these characteristics. But, if the burgeoning field of "complexity theory" has taught us anything, it is that matters cannot always be considered simple. Complex systems exist, so why shouldn't complex chance? In fact, as regards living systems, it seems fair to assert that complexity is more the rule than the exception. Are complex chance events to be precluded from the discourse on nature, as if they don't exist, just because they don't conform to known methods for measuring and regularizing?

Rogue Chance?

Our discovery is quite simple but a bit startling at first encounter: We are unable to encompass true qualitative change within the description of nature because we have turned a blind eye toward the existence of complex chance events. But exactly how common are complex chance events? Are we talking about the miraculous or something quite mundane? Physicist Walter Elsasser (1969, 1981) devoted significant attention to this question and came to the conclusion that complex chance events prevail everywhere there are living systems. More surprising still, he implied that they perfuse nature and even overwhelm the number of simple events by comparison.

One is tempted at first to dismiss Elsasser's claims, like those of Bateson, as preposterous, were Elsasser's assertions not so easy to rationalize and quantify. Elsasser treated the discontinuous nature of events in biology from two different perspectives. In the first, he considers complex events in terms of combinations of distinguishable types. All mathematicians know that the joint probability of several types randomly co-occurring diminishes rapidly as the number of types in the conjunction grows: if type A has a 50% probability of occurring and type B a 40% chance of independently happening, then the joint probability of both A and B randomly co-occurring becomes 20%. Now add types C, D, E, F, G, and H, all with roughly a 50% probability of independently appearing, and the joint probability that all eight types will randomly occur together falls to about 0.4%.

Elsasser then asks how many types would have to randomly co-occur before one could say with all reasonable certainty that any particular combination would never again recur by chance. In effect, he was searching for an upper boundary on physical reality, and so he asked what is the maximal number of simple events that possibly could have occurred throughout the history and extent of our known universe. Most recent estimates agree that there are about 10^{81} simple particles throughout all of known space. (That's 10 multiplied by itself 81 times).[1] Now the simplest physical events we can observe would happen to the simplest of particles over an interval that is characteristic of subatomic events (about a nanosecond or a billionth of a second). Because the universe has been around for some 13–15 billion years, or about 10^{25} nanoseconds, Elsasser, therefore, concluded that *at the very most* 10^{81} x 10^{25}, or 10^{106} simple events could have transpired. One can safely conclude that anything with less than one in 10^{106} chances of reoccurring simply is never going to do so, even over many repetitions of the lifetime of our universe. The take-home lesson is that one should be very wary whenever one encounters any number greater than 10^{106} or smaller than 10^{-106} because such frequencies simply cannot apply to any known physical reality. Elsasser calls any number exceeding 10^{106} an *enormous* number.

But exactly how does Elsasser's warning pertain to complex systems? The answer has to do with uniqueness. In particular, one asks how many different types or characteristics are required before a random combination can indisputably be considered unique. It may surprise some that this threshold is not an extremely large number. It is not Avogadro's number (roughly 10^{23}, the number of molecules in a mole of any chemical). It is not one million types, or even one thousand. Reliable uniqueness happens to require only about seventy-five distinct tokens because the combinations of types scale roughly as the factorial of their number. Because $75! \approx 10^{106}$, whenever more than seventy-five distinguishable events co-occur by chance, one can be certain that they will never randomly do so again.[2]

Elsasser's result is important to ecologists because it is almost impossible for anyone dealing with real ecosystems to consider one that is composed of fewer than seventy-five distinguishable individuals (e.g., Kolasa and Pickett 1991). For example, an ecosystem comprised of ten species, each represented by forty organisms would yield four hundred distinguishable entities.[3] One could argue that the conditional probabilities of co-occurrence in most systems would increase the likelihood of reoccurrence (and thereby raise the threshold for uniqueness). That is, given A, there is a good probability that B will also occur. This is true, but it doesn't buy one much, given the very finite lifetimes of unique biological entities. If, for example, one were to stand on the mezzanine at Grand Central Station and take a photograph of a twenty-meter square of the floor space below, one might count ninety individuals in the picture. Many of those individuals are commuters—creatures of habit who tend to pass the same place more or less at the same time each day (Figure 3.1). Nonetheless, the chances of capturing exactly the same ninety persons in a subsequent picture, before, say, half of them expire, is simply nil.

One concludes from Elsasser's calculations that singular

Figure 3.1. "Pedestrians—The Airport"

events are not rare; rather, they are legion! They occur everywhere, all the time, and at all scales! That last phrase is especially important because the works of Boltzmann and Fisher have put us in the habit of relegating chance events to only the microscopic realm. But we can now reason that uniqueness is even *more likely* at larger scales (where more complicated, and therefore readily distinguishable, entities exist). In fact, it is easy to construct a combinatorics argument similar to the one used by Elsasser to demonstrate that unique events and unique entities numerically overwhelm simple ones over all scales that are characteristic of life forms.

Opening the Window to True Change

The implications of Elsasser's warning are serious, indeed. As mentioned, in order to estimate a legitimate probability of an event, that event must reoccur at least several times. If an event is unique for all time, it evades treatment by probability theory. Now if the density of unique events overwhelms that of simple ones, as it does in complex systems, then most of reality lies beyond the ken of probability theory. Up to now, we have regarded chance as an exception, but we suddenly realize that it predominates in a complex world. The time has come to acknowledge explicitly the existence of radical contingency. Accordingly, let us formulate the first postulate of our new metaphysic as follows:

I. The operation of any system is vulnerable to disruption by chance events.

Had we remained unaware of singular events, the postulate would have seemed mundane. But Elsasser has robbed us of our innocence, and we now realize that the statement also pertains to the more pervasive actions of radical, singular occurrences that evade even rudimentary methods of prediction.

The postulate is also meant to affirm the ontic nature of chance. That is, chance is not merely an illusion to be explained away by the operation of laws, if only one knew matters surrounding the event with sufficient precision. To understand why one must foreswear universal lawfulness, it becomes necessary to know more about the nature of physical laws, as we will discuss later in chapter 6. To the contrary, the postulate is implying that the world is not a seamless continuum. The fabric of causality is porous. In analogy with Carathéodory's (1909) statement of the second law of thermodynamics, we can say that arbitrarily close to any event that is amenable to lawful resolution lie innumerable instances of radical chance (holes). Put in other words, the universe is not causally closed, but open in the sense of Popper (1982) and Peirce (1892).

Law and Mechanism in Biology—Absurdities?

By this point, the outlook for science is beginning to look rather grim, and it will become even worse before it gets better. (But the reader should rest assured that things *will* get better.) It seems for now that, in our quest to open science up to true change, we have allowed everything once comfortably inside its confines to escape. A world of continuous interruption is hardly amenable to any sort of simplifying generalizations, save the most trivial. In particular, we are questioning the explanatory power of laws in the world that immediately surrounds us. But lest we get carried away, I would hasten to add that no one is proposing that the physical laws are necessarily violated. To be clear about it, the proposal at hand is simply that physical laws are incapable of *determining* what we see in the living realm—that the combinatorics of complexity simply create so many possibilities, or degrees of freedom, that any physical laws can be satisfied in a vast multiplicity of ways. Another way of saying the same thing is that the realm of biology is *underdetermined* by physical constraints. (See also Kauffman 2000.)

To delve deeper into the insufficiency of physical laws, we turn to Elsasser's second argument as to why they cannot apply to biology. Elsasser stresses the heterogeneity inherent in biological systems. He notes that, in physics, one always deals with a continuum, whereas, in biology, the dominant concept is that of a class (such as a taxonomic species or an ontogenetic stage). Concerning the continuum assumption, Whitehead and Russell (1913) demonstrated that any particular treatment of a continuum can be logically mapped to an equivalent operation between homogeneous sets. A homogeneous set is a collection of tokens that are totally indistinguishable from each other. Examples of perfectly homogeneous classes would be a collection of electrons or hydrogen atoms. We have no means available to us of distinguishing one electron from another. More specifically, Whitehead and Russell proved that *lawful* behavior within the continuum can correspond *only* to operations between perfectly homogeneous sets.

While the rigorous logic used by Whitehead and Russell remains beyond the scope of this text, the underlying ideas can be conveyed via two very simple illustrations. We consider, for example, a collection of homogeneous sets of integers as follows: The first set consists of five tokens of the integer 1, the second contains five tokens of the integer 2, the third contains 3's, etc. Now we let the set of 2's interact with the set of 4's according to some strict operation. For example, each of the tokens in the first set might be multiplied by a corresponding member of the second. The result would be another homogeneous set of five eights (Figure 3.2). The *determinate* result is another single homogeneous set.

Now we consider collections of integers grouped by 5's according to magnitude. That is, the first group contains the integers 1 through 5; the second, 6 through 10; the third, 11 through 15, etc. Each of these aggregates is inhomogeneous; its members are clearly different each from the other. Now we let the first group operate on itself according to the same procedure we used in the first example. One possible result would be the integers 4, 5, 6, 8, and 15 (Figure 3.3). Noteworthy is that these products are scattered across three separate classes. Other combinations would yield similar *indeterminate* results in that they would scatter among several groups.

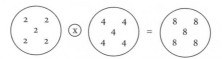

Figure 3.2. The result of a fixed operation upon two homogeneous sets. The result is a single homogeneous set.

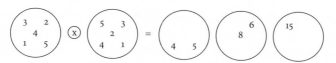

Figure 3.3. The same operation as in Figure 3.2 carried out between two heterogeneous groups of integers yields results that scatter across several different classes.

An example closer to biology might be a collection of alleles (genetic elements) with dominant character A and alternative form a. We start with a set of six pairings of these alleles, all with at least one dominant gene, A, say, AA, aA, Aa, AA, Aa, aA. Because in each instance it is the dominant character that is expressed, we might see no difference in the phenotypes (organisms) and assign them to a homogeneous group. Now we let these pairings couple randomly as in meiosis. One possible result might be AA, aa, Aa. Suddenly, we notice recessive trait aa for the first time and distinguish it as qualitatively different (possibly even a pathology or deformity). We recognize that there was hidden heterogeneity in our first grouping and so must create a second grouping to receive the aa progeny.

Biology, as Elsasser and so many others have shown, is rife with heterogeneity. He emphasized how Roger Williams (1956) had illustrated just how rampant heterogeneity and individuality were. Biotic entities can be distinguished all the way down the hierarchy. Even microbiota, usually considered indistinguishable, can be individuated. Of course, it often serves legitimate ends to ignore some differences and group into classes, but one should always remain aware that these are explicitly *heterogeneous* groupings (Krivov and Ulanowicz 2003). The members within any such class could always be subdivided, when and if necessary.

Citing the equivalence proved by Whitehead and Russell, Elsasser concluded that one cannot formulate a law in the Newtonian sense that would relate to operations among heterogeneous biological classes. (The possibility that universal laws are foreign to biology has been echoed independently by others [e.g., Mayr 1997, Lewontin 2000, Konopka 2007, Fox-Keller 2007].) It is difficult to overstate the importance of Elsasser's assertion because so much effort has been and is still being expended by those studying "ensemble" living systems (ecology, sociology, economics, etc.) to formulate laws for their fields— again the misguided "physics envy."

All that Elsasser has been telling us about both chance and heterogeneity can neatly be summarized in a few words: "Combinatorics and heterogeneity overwhelm law." That's not to say that any laws are broken. Rather, for any individual constraint, such as the conservation of energy in a particular situation, there is a multiplicity, and usually a great redundancy, in the number of patterns that could satisfy that law exactly. It follows that the law in question is incapable of differentiating among the many possibilities. Something else must intervene to determine the actual outcome, and that something will be described in the next chapter.

Because lawful behavior in physics fits what we generally regard as mechanical, it follows as a corollary of Elsasser's result that *the notion of mechanism in biology is devoid of logical underpinnings*. This certainly will come as news to many ecologists, for perusal of most texts, papers, and research proposals in ecology reveals a popular consensus that ecology is mostly about discovering and describing mechanisms. It is necessary that one opens one's eyes to the fact that the use of *mechanism* in ecology is metaphorical at best, and that a very real danger exists that the use of this metaphor can too easily divert one's attention in the *wrong* direction.

Eschewing Force in Biology

That *mechanism* in ecology (and in most of biology) is illogical, and at best metaphorical, might hardly qualify as news, even to those who dote on using the term. To be fair to my fellow ecologists, most are at least subconsciously aware that the metaphor is *far* from perfect. There is simply too much noise in ecological systems to accept the analogy literally. But why be satisfied with terminology that leads us astray of the true nature of reality? What if a more appropriate term for the interaction of two heterogeneous classes exists and conveys a clearer picture of the constraints actually at work?

The use of inaccurate language in science seriously troubled philosopher Karl Popper (1990). Unlike Elsasser, Popper did not tackle the issue of "law' directly. Rather, he questioned the generality of the related notion of "force" and whether it could be applied to biotic phenomena. Forces, Popper reasoned, exist only under idealized, isolated conditions, such as a vacuum. Under such conditions, whenever we encounter "cause" A, the result is perforce B. If A, then B; if A, then B; if A, then B; *ad infinitum*. Such is hardly ever the case in biology, however. True, most of the time that A occurs, it is followed by B—but not always. Quite often something else would interfere. (See Figure 2.3, for example.) Popper's example was the fall of an apple. The apple falls under gravity, but "Real apples are emphatically not Newtonian apples!" he declared. When an apple will fall depends not only upon its Newtonian weight but also upon the blowing wind, biochemical events that weaken the stem, and so forth. Exactly what happens and when it happens is conditional upon any number of other events or *interferences*.

As a result, Popper felt it was wrong to stretch the narrow notion of force to pertain to complex situations, where (again!) it could possibly lead one astray from what was really happening. He suggested, rather, that one speak of the *propensity* for an event to occur. If A happens, there is a propensity for B to occur, but B need not follow each and every time. The situation then becomes more like "If A, then B; if A, then B; if A, then B; if A, then C(!); if A, then B, etc." The nature of propensities can be elaborated with the help of an events table (Ulanowicz 1997).

An events table is a matrix array that shows how many times a putative "cause" A was followed by one of several possible results, B. One might label the causes $a_1, a_2, a_3, \ldots, a^m$. Similarly, one may cite a list of observed "effects," say $b_1, b_2, b_3, \ldots, b_n$. One then creates a matrix such that the entry in row i and column j shows the number of times (events) where a_i is followed immediately by b_j. The a_i, for example, might signify the ith type of human behavior among a list of m activities (e.g., eating raw meat, running, smoking, drinking alcohol). The b_j could denote the jth type among n kinds of cancer, such as that of the lung, stomach, pancreas, etc. The resulting events matrix would be filled with the numbers of individuals who habitually participated in the ith behavior and thereafter exhibited the jth form of cancer. Table 3.1 shows the hypothetical number of times that each of four causes was followed by one of five outcomes. For the sake of convenience, exactly one thousand events were tabulated. In an ecological context, the a_i in the table might represent various prey items that are fed upon by the predators b_j. In that case, the columns of the table would represent the diets of the respective predators. For example, if b_3 were to represent a predatory fish like the striped bass in Chesapeake Bay, its diet over a given interval might include sixteen organisms of blue crabs (a_1), thirty-eight individual silversides (a_3), and 161 menhaden (a_4). One would then say that striped bass have a propensity to consume menhaden, but, when other opportunities arise, they will also eat crabs and silversides.

Listed in the sixth column are the sums of the respective rows. Thus, a_1 was observed a total of 269 times; a_2, 227 times; etc. Similarly, the entries in the fifth row contain the sums of

	b_1	b_2	b_3	b_4	b_5	Sum
a_1	40	193	16	11	9	269
a_2	18	7	0	27	175	227
a_3	104	0	38	118	3	263
a_4	4	6	161	20	50	241
Sum	166	206	215	176	237	1000

Table 3.1. Frequency table of the hypothetical number of joint occurrences that four "causes" ($a_1 \ldots a_4$) were followed by five "effects" ($b_1 \ldots b_5$)

their respective columns. Effect b_4 was observed 176 times; b_5, 237 times; etc.

We see from the events table that the phenomena are not the result of determinate forces. Most of the time a_1 gives rise to b_2, a_2 to b_5, and a_4 to b_3. But there is also a lot of what Popper calls "interferences"—situations like those in which a_4 yielded b_1, which were occasioned either by some interjected external agency or by the interplay of processes within the system. One also notices that there is significant ambiguity as to whether the outcome of a_3 will be b_1 or b_4.

Although Popper focused exclusively on interferences, we now have the advantage of knowing Elsasser's opinion on the matter, and we immediately recognize that the same indeterminacy could have arisen because of heterogeneity. As with our examples of groupings of different integers, a single operation on a heterogeneous collection often gives rise to results that scatter across disparate groupings. Outcomes like those depicted in Table 3.1 should be quite familiar to those who deal in genetics, and such dispersion is the bane of those who conduct clinical trials in medicine.

Popper goes on to suggest that, if it were possible to isolate individual processes and study them in laboratory-like situations, then something resembling mechanical behavior might result. For example, we might discover that, if we take great care to isolate processes, a_1 always yields b_2, a_2 gives b_5, a_3 invariably results in b_1, and a_4 in b_3. (For example, a laboratory feeding experiment might be designed so as to pair single prey items with individual predators [e.g., striped bass have access only to menhaden.] As with most experimental configurations, such separation would constitute a highly artificial situation.) The same frequency counts taken from a collection of isolated processes might look something like those in Table 3.2.

Knowing a_i immediately reveals the outcome b_j in mechanical, lockstep fashion. What is also interesting in Table 3.2 is that

	b_1	b_2	b_3	b_4	b_5	Sum
a_1	0	269	0	0	0	269
a_2	0	0	0	0	227	227
a_3	263	0	0	0	0	263
a_4	0	0	241	0	0	241
Sum	263	269	241	0	227	1000

Table 3.2. Frequency table as in Table 3.1, except that care was taken to isolate causes from each other.

b_4 is never the outcome of an isolated cause. One surmises that, in the natural ensemble, b_4 is purely the result of interaction phenomena.

We note here for later reference that the transition from Table 3.1 to the configuration in Table 3.2 represents what happens when *additional constraints* are imposed upon the system being described. Noteworthy also is that values of the average mutual information (to be discussed) for systems with internal distributions like those in Table 3.2 are near or at their maxima.

We see how one may regard propensities to arise out of stochastic interferences upon mechanical forces, as Popper suggested, or to crop up because of heterogeneities *a la* Elsasser. In either view, propensities represent constraints, albeit imperfect ones, capable of holding systems together. Neither Popper nor Elsasser was very explicit as to how such constraints might originate and grow, but we will attempt a scenario in the next chapter. Suffice it for now to assume that propensities impart adequate coherence to a system to keep them from immediately disintegrating when impacted by most arbitrary singular events.

The Nature of the Fabric of Nature

And so matters are beginning to look up. We have encountered our first example of cohesion in the extra-Newtonian world.

For a long while after Newton, nature had been regarded as one seamless continuum. No holes or breaks in the continuum were conceivable. Furthermore, the edifice was rigid, like the proverbial clockwork. Movement anywhere propagated everywhere. With the discovery of quantum phenomena, many began to question the causal continuum at extreme microscales, but macroscopic phenomena were assumed to carry on with ineluctable Newtonian regularity. Our new conviction that singular events surround us at *all scales* forces us to replace the continuum with quite another image. Earlier we spoke metaphorically of the "fabric" of causality that holds the universe together. Between the threads of the fabric (propensities), we discern small holes (chance events) of various sizes. Instead of a fabric, some might prefer the three-dimensional image of a sponge, with its fractal-like distribution of holes of various sizes. Whichever image one might adopt, it should exhibit a porosity that lends a degree of flexibility or malleability because such openness is absolutely essential if evolution and/or development are to proceed.

One advantage of the Newtonian metaphysic was that it left no mystery as to what holds the world together. The laws of nature and their attendant forces were considered ubiquitous, eternal, and without beginning. But propensities are nowhere near as commanding as forces. How might they arise, and why do some come to dominate systems while others remain subsidiary? In other words, how can propensities grow in strength? One possibility emerges out of our shift from law toward process. Like Darwin, we seek some example of a natural process, or a configuration of processes, that could account for the emergence of order in the living world and its maintenance in the face of chance singular events.

4

How Can Things Persist?

*The predominant aim within the organism is survival
for its own coordinated individual expressiveness.*
—Alfred North Whitehead, *Modes of Thought*

Reversing Course

Our discussion up to now has been dominated by deconstruction, i.e., we have attempted postmodern-like arguments against
some enduring and formerly effective positions. If we are to make
a true contribution to understanding, however, it becomes necessary to adopt more of a constructivist outlook. If this sounds like a
sudden turnabout, then so be it, for the history of ideas has often
been punctuated by abrupt reversals in direction. John Haught
(2001), for example, documents the sudden turnaround in popular perceptions of life and death ushered in by the Enlightenment. Before the seventeenth century, life had been regarded as
ubiquitous and ascendant. It was seen to be present everywhere,
even in what today is perceived as purely physical phenomena.
The chief intellectual challenge for pre-Enlightenment philosophers, therefore, was to explain the exceptional nature of death.

Following Newton, however, the pendulum swung suddenly in the opposite direction. The preponderance of the universe now is considered to consist of dead, quiescent matter
that behaves according to deterministic and inexorable laws,

which leave no room for the irreversible, asymmetric, and contingent phenomena associated with living systems. Under such a vision, it should come as no surprise that one of the principal scientific and philosophical questions of our day has become the emergence of life: how could it possibly have arisen out of such a dead universe?

Our task now is to avoid either of these extremes and achieve some semblance of balance. In an effort to show how true change could be introduced into post-Newtonian thinking, we have virtually flooded the stage with radical chance events that violate the assumptions normally invoked to bring the aleatoric under reasonable control. True, toward the end of chapter 3, we did introduce the notion of propensity and suggested it as a counterbalance to the veritable flood of chance disturbances. But the vagueness of that argument doesn't make it too convincing. A propensity by itself doesn't seem robust enough to stem the tide. Popper (1990), however, insisted that propensities never exist alone but always stand in relationship to other propensities. We ask, therefore, whether the *juxtaposition* of propensities might possibly serve as an appropriate counterweight to the ubiquity of radical chance.

The World of Stasis

As a prelude to investigating the possibilities of juxtaposition of propensities, it would first behoove us to consider modes of persistence already familiar to us. Perhaps the predominant image is that of passive *equilibrium*. A crystal, a stone, Thomas Jefferson's desk, and the Parthenon are all solid objects that have been around for some time. They represent forms of matter that have come to thermodynamic equilibrium and will remain in that state until they are impacted by forces of considerable strength sufficient to pull them apart. The stability and endurance of material, especially when it persists well beyond the lifetimes of humans, have so impressed us that many see in material the most fundamental forms possible. Democritus and Lucretius imagined fundamental, simple atoms of solid mass as constituting building blocks from which everything sensible is composed—a world of "billiard balls and vacuum."

Passing the Torch

Actually, it is incorrect to say that equilibrium bears no relation to living systems because many biological systems appear to operate very near to equilibrium (Gladyshev 1997, Ho 1993). Ever since Darwin, however, we have become more aware of the interplay between change and persistence, although, even within biology, there has been a retrograde swing back to the realm of billiard balls. This is because the mode of endurance in the living world has become the genome, which, in turn, has been given particular didactic form in the DNA molecule. (Most biologists can probably hearken back to their introductory course and recall the plastic models of interlocking billiard-like spheres that were used to represent the DNA molecule.) The sequence of nucleotides in this polymer can be thought of as a string of code. These strings are passed from generation to generation and provide the material basis upon which a host of biochemical and enzymatic reactions can act in order to form offspring that will resemble their progenitors. So enamored of the many properties of DNA was one of its discoverers that he could not imagine it to be of terrestrial origin (Crick and Orgel 1973). Others believe it unlikely that naked DNA emerged from purely physical surroundings and later was dressed with the "machinery" that reads it (Kauffman 1995). More probably, DNA evolved within a matrix of proto-organisms that were already capable of growth and reproduction before the advent of this specialized molecule (Deacon 2006). Despite this relatively passive role, most researchers today regard DNA as efficient cause and count as secondary the constellation of biochemical processes surrounding it, whose progenitors likely brought the molecule into existence.

Actively Rendering Order out of Chaos

Kauffman, Deacon, I, and others are driving at the ostensible paradox that out of a mélange of processes can emerge certain patterns of transformations that endure over time. For how else could the hard material of this world have arisen? We could believe, as did the early Platonists, that matter arose to conform to preexisting "essences," but contemporary physics paints a considerably different picture of what we regard as hard material. For one, what to our senses seems solid, we are told, is, in large measure, empty space. And so we take recourse in assuring ourselves that the elementary particles found within that space have been around since the beginning of time—well, almost the beginning of time, that is. According to the theory of the big bang, the universe began as a chaotic, incredibly dense mass of extremely high-energy photons—pure flux (Chaisson 2001). As this continuum began to expand, some of the photons came together (collided) to form pairs of closed-looped circulations of energy called *hadrons*, the initial matter and antimatter. For a while, these hadrons were destroyed by collisions with photons about as fast as they appeared. Continued expansion put space between the elementary particles so that matter and antimatter pairs annihilated each other with decreasing frequency, and the diminishing energy of the photons made their collisions with extant material less destructive. Matter was beginning to appear but was also disappearing at much the same rate.

Meanwhile, a very subtle (one in a billion) asymmetry (chance event) produced slightly more matter than antimatter, so that, after most antimatter had been annihilated by matter, a remainder of matter slowly accrued. Further expansion gave rise to yet larger configurations of emerging materials and the appearance of weaker forces. Eventually matter coalesced under gravity (in stars) to a density that ignited chain (feedback) fusion reactions, producing larger, more complex aggregations—the heavier elements. From these, it became possible to construct solid matter. The take-home message here is that the enduring materials we perceive today are actually the endpoints of dynamical *configurations of processes*, asymmetries, and feedbacks of bygone eons.

In the light of this expanded view of the universe afforded by cosmology, we again ask where in this concoction might lie the key to rendering order out of chaos. Recognizing that stationary forms are subsequent to movements and processes, the question could be rephrased as the following: what process or combination of processes might yield ordered form out of chaotic substrate?

I believe the definitive clue to answering this question was provided in a remark made by our protagonist, Gregory Bateson (1972, 404): "In principle, then, a causal circuit will generate a non-random response to a random event."[1] Bateson, the cyberneticist, was dealing with "causal circuits," concatenations of events or *processes* wherein the last element in the chain affects the first—what commonly is known as *feedback*. Causal circuits, he implied, have the capability to endure because they can react nonrandomly to random stimuli. This inchoate fragment of an idea, I believe, is the keystone in our efforts to understand the life process and its origins. It is the basis for what resolves the outcome whenever multiplicity has overwhelmed physical law. Accordingly, I choose the following as our second fundamental assumption:

II. A process, via mediation by other processes, may be capable of influencing itself.

Now, most readers will immediately object that this is nothing particularly new. It is the description of feedback, a phenomenon familiar to almost everyone and an element of conventional science. And so it is. Others will offer that it is nothing more than the recognition that nonlinearities exist in the world. This is true as well. What, then, is so extraordinary about the statement that would warrant choosing it as a fundamental postulate of how nature works? It is, after all, possible to study feedback within the ruins of the Newtonian construct quite nicely, thank you.

The problem with feedback, however, is that it is difficult for most of us to approach the subject from outside of its own legacy: feedback rose to prominence in science with the advent of the cybernetic movement in the 1940s. In its earliest forms, it was a dynamical element of artificial systems such as radio circuits or control circuits, which had been designed and fabricated in atomistic fashion. Only later was the concept abstracted from human design and applied to the natural. Even there it has remained implicitly coupled with the notion of atomism, and most of us were thoroughly inculcated with atomism and decomposability during the course of our formal training. Thus situated and supported, feedback occupies a very legitimate place in contemporary science.

I hope it will not ruin the reader's suspense if I telescope one of my fundamental suggestions at this point—namely, that I am going to recommend that we drop entirely the Newtonian assumption of atomism as a key postulate. As I hope to demonstrate, the assumption simply does not accord with contemporary phenomenology and invoking it leads us astray from the essence of complexity in nature—but more about that later. What I wish to emphasize now is that to acknowledge feedback in complete abstraction from atomism radically changes our view on nature.

Whenever feedback acts within a system that has been assembled from parts and that can be disassembled into pieces that continue to maintain their separate identities and behaviors, it can be explained as the result of bottom-up causalities (neglecting, of course, any intentionality of the part of the builder, who, anyhow, would be disregarded as an epiphenomenon by those clinging to atomism). Banish atomism, however, and the postulate can be (and is intended to be) interpreted as admitting that causal action may arise at the level of observation. At the same time, the second postulate, without any auxiliary atomism, overtly extirpates the classical assumption of closure. The new axiom is a relational one. A process engages in interaction with other processes and affects itself as a result. The configuration of processes is affecting its members—a mereological notion that bears shades of top-down or Aristotelian final causality—the absolute bane of Enlightenment thinkers.

At this point, some readers might be tempted to close this book for good and return to other activities. Thus do I hasten to assure them that embracing feedback and dispensing with atomism does not constitute a flight into transcendentalism. On the contrary, it is a step toward bringing rationality back into our approach to living systems. Those who traffic in feedbacks are acutely aware that the phenomenon incorporates circularity and that logic abhors a circular argument. Thus are cyberneticists always at risk of committing circular reasoning and must indulge in all manner of circumlocution to avoid sounding illogical. By gathering all feedback into a single postulate, however, one excises circularity with one fell swoop from all subsequent arguments. Circularity becomes a given—a self-evident element of existence. With this conceptual restructuring now behind us, I beg the reader to remain with the discourse to see what can be logically and profitably constructed upon this new foundation.

To start with, the literature on feedback and cybernetics is immense, and there is no way to do justice to it all here. Suffice it to remark that the early approaches to feedback all involved linear or quasilinear mathematics. For a long while, this confined research exclusively to negative feedback because positive feedback in linear systems is pathologically destabilizing. Any action that amplifies itself in a linear way grows without bounds.

Studying only negative feedback was not initially debilitating because most applications of feedback pertained to the control of fabricated systems, where negative feedback is desirable. For example, the common governor was built into steam and internal combustion engines to control the speed of their operation. The governor was usually some form of centrifugal weights that were spun by the engine. As the engine accelerated, the weights moved further from the center of rotation and were levered to a valve that decreased the supply of steam or fuel to the engine, thereby slowing the increase. If the rotating weights had been levered so as to supply *additional* fuel as speed increased (positive feedback), the engine would have raced out of control and destroyed itself. In a linear world, positive feedback is always destructive.

Once the study of feedback moved beyond artificial contrivances and into the nonlinear, highly dissipative realm of living systems, it became necessary to reconsider the role of positive feedback. Living systems dissipate a large fraction of what they ingest, so that some sort of stimulation is required to keep them operating. Because organisms are largely self-entailing (Rosen 2000), the most reasonable place to search for such stimulation was within the system itself. Thus it was that any number of biologists came to investigate the role that positive feedback (autostimulation) plays in the life process (e.g., Eigen 1971; Haken 1988; Maturana and Varela 1980; Kauffman 1995; DeAngelis, Post, and Travis 1986; Ulanowicz 1986; and Deacon 2006). Much of that focus has come to dwell on autocatalysis as the genesis of both form and stability in evolving systems (Ulanowicz 1997, Kauffman 1995).[2]

Autocatalysis as Dynamical Agency

Autocatalysis is a particular form of positive feedback wherein the effect of every consecutive link in the feedback loop is positive. Such facilitation need not be assumed obligate and rigid, as with mechanical systems. There simply needs be present the *propensity* for each participant to facilitate its downstream member. As inferred above, whenever autocatalysis acts in tandem with atomism (as is legitimate when applied to most simple chemical systems), it may be regarded as wholly mechanical in nature. However, in more complex situations, as concern us here, and especially in conjunction with complex chance events, several distinctly nonmechanical attributes suddenly appear.

The nonmechanical attributes of autocatalysis become more apparent once they are illustrated in terms of a particular but generic example. Without loss of generality, I direct the reader's attention to the simple three-component interaction depicted in Figure 4.1. We assume that the action of process A has a propensity to augment a second process B. (To repeat for emphasis: A and B are not tightly and mechanically linked. Rather, when process A increases in magnitude, most [but not all] of the time, B also will increase.) B in its turn tends to accelerate C in similar fashion, and C has the same effect upon A.

My favorite ecological example of autocatalysis is the community that builds upon the aquatic macrophyte, *Utricularia* (common name "Bladderwort." See Ulanowicz 1995b). All members of the genus *Utricularia* are carnivorous plants that trap and ingest animals, like the familiar Venus Fly Trap, except *Utricularia* traps its prey using small utricles or bladders scattered along its feather-like stems and leaves (Figure 4.2a). Each utricle

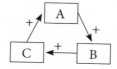

Figure 4.1. Schematic of a hypothetical three-component autocatalytic cycle.

has a few hair-like triggers at its terminal end, which, when touched by a feeding zooplankter, opens the end of the bladder and the animal is sucked into the utricle by a negative osmotic pressure that the plant had maintained inside the bladder. In the field the surface of *Utricularia* plants always supports a film of algal growth known as periphyton (Figure 4.2b). This periphyton serves in turn as food for any number of species of small zooplankton. The autocatalytic cycle is completed when the *Utricularia* captures and absorbs many of the zooplankton. To summarize, the growth of *Utricularia* provides an expanding platform upon which more periphyton can grow. More

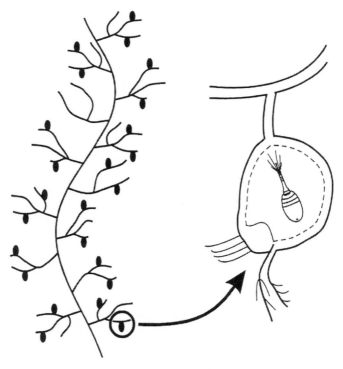

Figure 4.2a. Sketch of a typical "leaf" of *Utricularia floridana,* with detail of the interior of a utricle containing a captured invertebrate.

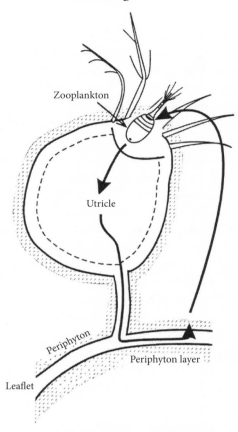

Figure 4.2b. Schematic of the autocatalytic loop in the *Utricularia* system. Macrophyte provides necessary surface upon which periphyton (speckled area) can grow. Zooplankton consumes periphyton, and is itself trapped in bladder and absorbed in turn by the *Utricularia*.

periphyton means more food for zooplankton, and more zooplankton result in more captures by *Utricularia*, which facilitates the latter's growth.

Generally, autocatalysis among propensities gives rise to at least eight system attributes, which, taken as a whole, comprise a decidedly nonmechanical dynamic (Ulanowicz 1997). For one, autocatalysis is explicitly *growth-enhancing* (by definition). Growth anywhere engenders growth everywhere else along the loop. Furthermore, autocatalysis exists as a *formal* structure or configuration of processes. Most germane to the discussion of evolution is the fact that autocatalysis is capable of exerting *selection* pressure upon its own ever-changing constituents. To see this, let us suppose that some small change occurs spontaneously in process B. If that change either makes B more sensitive to A or a more effective catalyst of C, then the change will receive enhanced stimulus from A. Conversely, if the change in B either makes it less sensitive to the effects of A or a weaker catalyst of C, then that change will likely receive diminished support from A. For example, if some change in the periphyton community were to make it more digestible or a better food to the zooplankton, one would expect the zooplankton to increase and yield more support for the growth of *Utricularia*, which, in turn, would provide more surface area for that type of periphyton to colonize. Conversely, a change in periphyton that renders it less digestible might benefit the attached algae in the short term but would decrease the growth of *Utricularia* over the long haul, and that periphyton would be faced with less available habitat. In this system, autocatalysis would select for more digestible periphyton.

Once more we reiterate that all linkages are contingent in nature, not deterministic. We further note that such selection works not only on the component objects but on the processes and propensities as well. Hence, any effort to explain the development of an autocatalytic configuration in terms of a fixed set of elements or processes (i.e., via causal reductionism) is doomed ultimately to fail. These features should help to make clear how atomism and closure could preclude from our vision the full range of autocatalytic behavior.

It is also important to note that the selection pressure exerted by autocatalysis arises from within the system. This is in stark contrast to Darwin's conscious effort to preserve the Newtonian framework by externalizing the agency of selection. The action of autocatalytic feedback tends to *import* the environment into the system or, alternatively, *embeds* the system into its environment.[3]

Here it becomes necessary to interrupt the discussion of the properties of autocatalysis to interject the third and final postulate upon which our narrative about developing systems will rest: because there are differences between the forms on which selection is acting, there must be some way for those forms to substantiate and retain their history throughout the course of that selection. Accordingly, we recognize as fundamental what Darwin brought into scientific discourse:

III. Systems differ from one another according to their history, some of which is recorded in their material configurations.

In these days of molecular biology, the last clause immediately conjures up images of DNA or similar polymers, and these indeed play a major role in how systems convey form and dynamics from one generation to the next. But "material configuration" is by no means limited to static modes of retention. It is necessary to think more broadly because there most assuredly were proto-organisms before DNA and their like ever evolved. Furthermore, the mode of recording doesn't even have to imprint upon a persistent object. History can endure as well through time as a very stable configuration of processes, which reestablishes itself whenever the system is disturbed. In many ways, the structure of activities within a society embodies the history of that society every bit as much or more than the aggregate DNA of the individuals that make up the community (cf. Wilson 1980).

Implicit in axiom III is the notion of information, or what Gregory Bateson (1972, 475) called "the difference that makes a difference." Because biological entities are heterogeneous, they possess more information than their nonliving counterparts, such as mass, energy, or atoms of simple molecules, which cannot be distinguished one from another. Furthermore, the laws of matter and energy are all reversible and, therefore, conservative (Noether 1983). The dynamics of information, however, resemble and complement the irreversible and nonconservative production of thermodynamic entropy. Although these dynamics of information do not violate the reversible laws of nature, neither are they fully restrained by them (e.g., chapter 6., Jørgensen et al. 2007). As Elsasser concluded, given sufficient complexity, the combinatorics of structures that contain information overwhelm physical laws. Bateson, for his part, implicitly recognized such extralawful behavior by distinguishing between two fundamental categories of existence—the "creatura" and "pleroma" mentioned earlier. Critically aligned with Bateson's distinction is the notion that history can serve as a criterion for identity.[4] In other words, the trajectory of a system through time conceivably could be used in lieu of a set of its existing properties.

Now knowledgeable about ways that systems can retain their histories, we can return to our catalogue of the assets of autocatalysis. In doing so, we immediately recognize a corollary feature of selection—namely, that it exhibits an extremely important behavior called *centripetality*. The centripetal nature of autocatalysis becomes evident as soon as we realize that any change in B is also likely to involve a change in the amounts of material and energy that flow to sustain B. In our *Utricularia* example, for instance, if the periphyton is starved for phosphorus and any change (or immigrant species) enables the film of algae to increase its activity by taking in more phosphorus, that change will be rewarded by the loop. From this, we perceive a tendency to reward and support those changes that bring ever more resources into B. As this circumstance pertains to all the other members of the feedback loop as well, any autocatalytic cycle becomes the center of a centripetal vortex, pulling as many resources as possible into its domain (Figure 4.3). Even in the absence of any spatial integument, the autocatalytic loop itself defines the locus of this organic trait. That is, centripetality becomes a core element of a system's identity.

Most readers are familiar with the propensity of living systems to sequester materials and energy. Anyone who has had to remove a beehive built within the wall of a dwelling has doubtless marveled at the sheer mass of honey and wax that had been accumulated. But such aggregations reflect the action of a particular part of an ecosystem, rather than the community as a whole. More to the point are tropical coral reefs, which, via their considerable synergetic activities, draw a richness of nutrients out of a desert-like and relatively inactive surround-

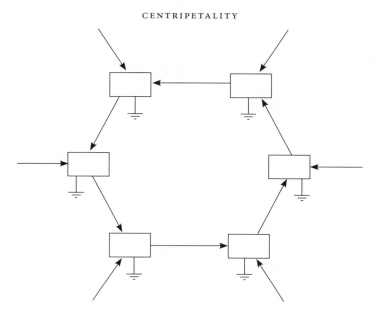

Figure 4.3. Centripetal action as engendered by autocatalysis.

ing sea. Then there are the soils that cover the floors of most terrestrial biomes. That this soil accumulates through the action of the entire community is best demonstrated by examples of the process in reverse. That is, when a community is sufficiently disturbed, it no longer is able to retain its soil. There are numerous examples of such atrophy—perhaps the most startling is the disappearance of the deep soils that once covered the island of Iceland.[5] Only centuries ago, before the advent of humans, the island was covered with heavy forest and lush soils (Hallsdottir 1995). The wood of the forest was soon exhausted building housing and ships, while the settlers let loose sheep they brought from Europe to graze upon the landscape (Kristinsson 1995). Today much of Iceland resembles a moonscape. In fact, parts of Iceland were used by NASA during the 1960s to practice for their moon missions. Loss of centripetality accompanied the disruption of community functioning, and with it went the system's legacy of soil.

It is very difficult to overstate the importance of centripetality. It is a largely neglected, but absolutely essential attribute of living systems. Furthermore, centripetality is an agency proper to the loop as a whole. Although the accumulation of resources is accomplished at the compartmental level, the drive to increase such activity is strictly a consequence of the relational structure of the whole. As mentioned in chapter 2 in connection with Darwin's theory, a very important but unstated premise of his scenario is that participants strive to capture and accumulate resources. The conventional Darwinian narrative does not mention the origins of this drive, but we now see it as the deductive consequence of autocatalytic action.

The renowned philosopher Bertrand Russell (1993, 22) was among the first to appreciate the central importance of centripetality to evolution, although he referred to it under the guise of "chemical imperialism":

Every living thing is a sort of imperialist, seeking to transform as much as possible of its environment into itself and its seed. . . . We may regard *the whole of evolution* as flowing from this "chemical imperialism" of living matter. (emphasis mine)

Unfortunately, he did not offer to explain the etiology of such "imperialism," but it was obvious in the way Russell phrased his comment that he was *not* referring to the disembodied, external action of "natural selection." Rather, he considered the striving, so conveniently ignored in almost all discussions on evolution, to be both innate and constitutional. Hence, although the origins of Russell's drive remain a tacit mystery under conventional Darwinism, here it becomes an explicit consequence of the second postulate of process ecology.

To underscore the fundamental and essential status that Russell accorded centripetality, we now assert that competition is derivative by comparison. That is, whenever two or more autocatalyic loops draw from the same pool of resources, it is their autocatalytic centripetality that *induces competition* between them. By way of example, we notice that, whenever two loops partially overlap, the outcome could be the exclusion of one of the loops. In Figure 4.4b, for example, element D is assumed to appear spontaneously in conjunction with A and

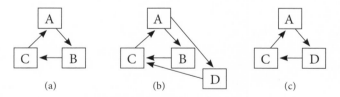

Figure 4.4a. Original configuration. *Figure 4.4b.* Competition between component B and a new component D, which is either more sensitive to catalysis by A or a better catalyst of C. *Figure 4.4c.* B is replaced by D, and the loop section A-B-C by that of A-D-C.

C. If D is more sensitive to A and/or a better catalyst of C, then there is a likelihood that the ensuing dynamics will so favor D over B, that B will either fade into the background or disappear altogether. Innate selection pressure and centripetality can actually guide the replacement of elements.

Ecology and evolutionary theory hew tightly to Darwin's example by separating factors of selection from the organisms on which they act. As a result, these fields have, to the best of my knowledge, produced no explicit example of autocatalytic replacement. Economists, however, being less constrained by Darwinian orthodoxy, seem more open to autocatalysis. Igor Matutinović (2005, 12), for example, writes:

Firms with the strongest autocatalytic loop experience the fastest growth and, consequently, considerably outgrow other competitors. Some of these firms—usually the early entrants—that choose to position themselves in the center of the market where they serve the widest range of needs are most likely to become the new industry hubs.

Returning to Figure 4.4, if B can be replaced by D, there remains no reason that C cannot be replaced by E and A by F, so that the cycle A, B, C could eventually transform into F, D, E (Figure 4.4c). One concludes that the characteristic lifetime of the autocatalytic form usually exceeds that of most of its constituents. This longevity is exceptional only in that it is sustained in absence of any external agency of repair or development. That is, dynamics remain wholly entailed within the system (Rosen 2000). By way of example, virtually none of the cells that composed our bodies seven years ago (with the exception of our neurons) remains as parts of us today. A very small fraction of the atoms in our body were in us eighteen months ago. Yet, if your mother were to see you for the first time in ten years, she would recognize you immediately.

So, we see how a configuration of processes can, as a whole, strongly affect which objects remain in a system and which pass from the scene. This observation inverts, to a degree, the conventional wisdom that it is objects that direct processes. The processes, as a union, make a palpable contribution toward the creation of their constituent elements. This reversal of causal influence lies at the crux of process ecology, and it extirpates the Newtonian stricture of closure (Ulanowicz 2001).

One should never lose sight of the fact that the autocatalytic scheme is predicated upon mutual beneficence or, more simply put, upon mutuality. Although facilitation in autocatalysis proceeds in only one direction (sense), its outcome is, nevertheless, mutual in the sense that an advantage anywhere in the autocatalytic circuit propagates so as to share that advantage with all other participants. That competition derives from mutuality and not vice versa represents an important inversion in the ontology of actions. The new ordination helps to clear up some matters. For example, competition has been absolutely central to Darwinian evolution, and that heavy emphasis has rendered the origins of cooperation and altruism obscure, at best. Of course, scenarios have been scripted that attempt to situate cooperative actions within the framework of competitions (e.g., Smith 1982). But these efforts at reconciliation invariably misplace mutuality in the scheme of things. Properly seen, it is the platform from which competition can launch: without mutuality at some lower level, competition at higher levels simply cannot occur. The reason one rodent is able to strive against its competitors is that any individual animal is a walking "orgy of mutual benefaction" (May 1981, 95) within itself. Alternatively, mutuality manifested at higher levels fosters competition at levels below, as we saw with the competition between B and D in Figure 4.4b.[6]

Brian Fath and Bernard Patten (1998) have suggested that, even if we remain entirely at the focal level (where competition is most apparent), there is reason to expect that matters will propagate through the network and evolve over time so that mutuality will gradually displace or overcome predatory and competitive interactions. This will happen whenever positive indirect effects grow to be more influential than direct negative effects. For example, during the wet season in the forested regions of the Everglades, alligators account for 10 percent of total predation on crawfish, and turtles account for double that amount. The alligator, however, preys on turtles as well. The effect of alligators eating turtles is beneficial to the crawfish because it reduces the amount of crawfish lost to turtles. In quantitative terms (Ulanowicz and Puccia 1990), the cumulative indirect effect that alligators have on crawfish is positive because alligators consume enough of the crawfish's other predators to more than compensate for the negative impact of their direct predation (Bondavalli and Ulanowicz 1999). Whether indirect effects between any two compartments are positive or negative is a function of the relative magnitudes of trophic exchanges along all pathways linking the two species (Ulanowicz and Puccia 1990). Fath and Patten maintain that these magnitudes are constantly adjusting in such a way that the positive indirect effects of the type just mentioned gradually become more prevalent.

That mutualistic configurations would grow and persist is no great mystery—autocatalytic mutualism is the only combination of interactions that *uniformly* supports the growth and continued existence of all members. Once a negative interaction enters at any point, that effect may propagate in ways that work to the detriment of at least some members. As we have seen, finite resources inevitably will lead to negative interactions at other levels, but their appearance is accidental (in the philosophical sense that they arise for extrinsic reasons). Mutuality, by contrast, is ontologically primal. It is an essential condition that merits foundational status.

A further feature of centripetality is that the drive it generates imparts an inherent direction to the system, namely, that of ever-more effective autocatalysis. In the terminology of physics, the unidirectional sense of autocatalysis is *symmetry-breaking*. I should hasten to add, however, that one should not confuse this rudimentary directionality with full-blown teleology. For example, it is not necessary that the system strive toward some preordained endpoint. The direction of the system at any one instant is defined by its state at that time, and that state changes as the system develops. This situation can perhaps best be envisioned as "localized hill-climbing,"[7] where the direction the system takes is set by its local neighborhood and remains insensitive to conditions farther away. I have used the term *telos* to denote this weaker form of directionality and to distinguish it from the rarer and more complex behavior known as teleology—Darwin's predilection for the latter notwithstanding.

Yet another important aspect of the asymmetry in auto-catalysis is that, whenever such a system is perturbed, auto-catalysis acts in homeostatic fashion to restore its own intensity. Sometimes this occurs so as to restore the system to how it appeared before disturbance. The system "heals" itself, so to speak. We might imagine, for example, element B in Figure 4.4 becoming defective and being replaced by a functioning homolog, B' (instead of D). In this manner, systems can be stabilized with respect to encounters with radical chance events of small magnitude. This same homeostasis, when viewed in hierarchical perspective, implies that the radius of disturbance of any chance event will remain circumscribed by autocatalytic selection at higher levels. In other words, the larger system is insulated, to a degree, from much of the disturbance acting at lower levels. Unlike the rigidity and brittleness of Newtonian systems, chance need not destroy an ecological system.

The combined actions of selection and asymmetry serve to fortify those propensities that contribute most to the autocatalytic action within the system. This answers the question posed toward the end of the last chapter as to how propensities grow in strength. If left unperturbed, autocatalysis would reinforce propensities to a degree that they begin to act in mechanical, law-like fashion. The effect of autocatalysis is to strengthen the "fabric of causality" and make it less flexible in the process. The overall trend is for the constraints in processes to start out weak and arbitrary, yielding outcomes that differ only marginally from the stochastic (see Figure 2.1). Gradually (or abruptly in rare cases), the constraints will continue to strengthen each other in the absence of major perturbations until their behavior becomes almost law-like (as in Figure 2.3). Such transition has been referred to as *canalization* (Waddington 1942), and we shall pursue such directionality further in the next section.

Taken together, selection pressure, centripetality, and a longer characteristic lifetime all give evidence that the larger organic structure begins to exhibit a degree of *autonomy* from its constituents. This should come as no big surprise, however, because we have discussed how such autonomy becomes implicit in the second axiom just as soon as one dispenses with atomism. One conclusion worth repeating, however, is that attempts at reducing the workings of an organic system exclusively to the properties of its composite elements always prove futile over the long run.

The organic dynamics just described can be considered *emergent* in the epistemic sense (i.e., virtual). Organic behavior all too often remains cryptic, however, due to a common predilection to view matters narrowly. In Figure 4.5, for example, if one were to consider only those elements in the lower right-hand corner (as enclosed by the solid line), one could readily identify an initial cause and a final effect. If, however, one were to expand the scope of observation to include a full autocatalytic cycle of processes (as enclosed by the dotted line), then the ensemble of system properties we have enumerated would become immediately evident. For example, if an ecologist were to concentrate on studying the consumption by zooplankton of periphyton attached to *Utricularia* in limestone lakes in Florida,

Enlarged system boundary

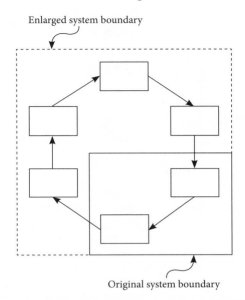

Original system boundary

Figure 4.5. Two hierarchical views of an autocatalytic loop. The original perspective (solid line) includes only part of the loop, which, therefore, appears to function rather mechanically. A broader vision encompasses the entire loop and, with it, several nonmechanical attributes.

it is unlikely he/she would ever discover why the *Utricularia* community is dominant in that particular habitat (Ulanowicz 1995b).

It is clear that such emergence is only a matter of appearances and perspective. A far more important issue is whether anything can truly emerge from a system that wasn't already hidden in it. That is, is emergence possible in the ontological sense of the word? There is good reason that ontic emergence was impossible under the Newtonian metaphysic: the restriction of closure precludes anything from arising that isn't already present at lower levels. Interestingly, Prigogine (1978) theorized how indeterminacies at microscales could cause shifts in behavior at the macrolevel; but the indeterminacies he allowed were always simple and generic, and the options for system change were few in number (usually only two) and already immanent in the existing system configuration (Ulanowicz 2005). Our recognition of complex, singular events, however, opens an avenue for the emergence of truly novel changes, whenever particular combinations of novel disturbances come together with the current system structure in lock-and-key fashion. Of course, in order to demonstrate that these dynamics truly emerge, one must show how they are capable of exerting influence on subsystems from which they emerged (Clayton 2004; Peterson, forthcoming). We have already addressed the feasibility of top-down influence, and a more thorough consideration of how emergent configurations can affect their constituents will follow in the next chapter.

Arriving at the Beginning

With a rather extensive catalog of the attributes of autocatalysis now elaborated, we are thus prepared to entertain the key concept that, for me, initiated the new perspective being developed here. As I mentioned in chapter 1, it was the chance discovery of a phenomenological formula that set things in motion. In 1978, it was my good fortune to happen to have read papers by two independent authors in close succession to one another. This fortuitous juxtaposition led me to formulate a single quantitative index to serve as a measure of a system's growth and development (Ulanowicz 1986).

The first article was Henri Atlan's (1974) argument on the existence of a threshold of complexity that a system must exceed before it can exhibit "self-organization." In other words, a system must attain a certain level of complexity before it can interact with its environment in a way that increases its own organization. In the course of his argument, Atlan had suggested that one use the average mutual information (AMI), a logarithmic index from the mathematical field of information theory, as a functional measure of "organization." In an organized system, most of its components communicate with only a restricted subset of other components, i.e., communication and power are channeled along preferred routes. Constraints that guide flow over restricted pathways contribute heavily to the AMI. Unfortunately, although Atlan provided a mathematical formula for the AMI, he left unclear how one could actually determine a value for the AMI in any given situation.

Only a day or two after reading Atlan, I serendipitously encountered a paper by Rutledge, Basorre, and Mulholland (1976), who also utilized the AMI in the process of estimating certain properties of quantified ecological networks, *sensu* Lindeman. The aim of these authors had been to quantify the potential for an ecosystem to endure, due to the presence of parallel connections within its web of interactions—that is, a system with parallel pathways linking the same components possesses strength-in-reserve. Such multiplicity of pathways allows a system to compensate for a disturbance in one of the routes by increasing its activity along those that parallel the impacted conduit (Odum 1953). For example, in Figure 4.6, there are three pathways from element A to component D. In the cypress wetlands of Florida, D might represent the American alligator, a top predator that consumes, among other things, snakes (B) and turtles (C). All three of these animals also consume crawfish (A) as part of their diets (Bondavalli and Ulanowicz 1999). Should some condition (such as a species-specific virus) arise that decimates the turtle populations (B), the overall flow of material and energy between crawfish and alligators could be maintained by compensatory flows in the direct consumption of crawfish by alligators (A→D) and/or the indirect consumption of crawfish as mediated by the snakes (A→C→D).

The reader may already have noticed that the concerns of

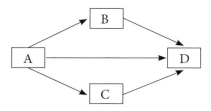

Figure 4.6. Three parallel pathways connect species A to species D.

Atlan and Rutledge and his associates are complementary or mutually exclusive. Such opposition is reflected in the way that Rutledge et al. calculated their measure of potential stability. They suggested an information index known as the "conditional entropy" to gauge the system-wide prevalence of parallel connections. The problem with the conditional entropy is that it is difficult to consider on its own (as suggested by the qualifier "conditional"). It can only be defined and calculated in relation to two other network attributes. The first of these was the overall diversity of system flows, which may serve as a surrogate for the overall complexity of the network. Over two decades earlier, Robert MacArthur (1955) had used the Shannon (1948) formula (often called the statistical entropy) from information theory for precisely this purpose. The second attribute was the organization of the flow structure, which Rutledge, Basorre, and Mulholland, like Atlan, also identified with the average mutual information (AMI).

The broader idea at work here was that complexity could appear in either organized or disorganized forms. MacArthur had already demonstrated how one can estimate the overall complexity, so, if it were possible to quantify how much of that complexity appears in organized form, then one could subtract the organized component from the overall estimate, to arrive at an estimate of the conditional entropy—a residual complexity that can be considered as disorganized or flexible. As a word equation, this operation would appear:

conditional entropy = overall complexity − organized complexity,

or, alternatively,

overall complexity = organized complexity + conditional entropy.

The latter relationship implies that complexity can be parsed into two distinct components: one that aggregates all the coherent constraints inherent in the system and a complement that pools all the disorganized and unencumbered complexity. The operation parses order from chaos, so to speak. Information theory is the only field of mathematics (known to me) that uses the same mathematical terminology to treat both constraint and indeterminacy.

(Some readers may still be wondering why parallelism contributes to the conditional entropy. It is because parallelism always generates indeterminacy. For example, it cannot be determined before the fact whether a quantum of carbon embodied in a crawfish will be conveyed directly to the stomach of an alligator or indirectly via a snake or a turtle.)

The genius of Rutledge and his associates lay in how they actually used the topology and magnitudes of the network flows to assign a number to the AMI, thereby opening the way to calculate the conditional entropy of any arbitrary, quantified network.[8] The Rutledge group then turned their attentions to quantifying the conditional entropies of various ecosystem flow networks to see if they could tie their results to the relative stabilities of these systems. With Atlan still fresh in my mind, however, my attention immediately was drawn instead to the intermediate term that Rutledge had used to calculate the conditional entropy, namely, the AMI. Atlan had showed how it quantifies the coherence among flows. I was excited to encounter an explicit method for how one can numerically assess the organization inherent in a quantified network of ecosystem transfers.

It bears repeating yet again that the relationship between

the AMI and the conditional entropy is complementary, that is, the two sum to equal the overall flow complexity (Figure 4.7). If this overall complexity should happen to hold nearly constant (for reasons we shall discuss presently), then any change in the AMI would have to take place at the expense of the conditional entropy and vice versa. That is, to the degree that complexity does not change, the two measures are agonistic and mutually exclusive: AMI tracks a system's organization, while the conditional entropy traces its relative *disorganization.*

Rutledge's methods of calculation opened the door to applying information theory to weighted ecosystem networks in a very practical way. Perhaps even more importantly, their work appeared to unite what Bateson (1972, 460) had called the two irreconcilable faces of ecology: (1) matter and energy, and (2) information:

Ecology has currently two faces to it: the face which is called bioenergetics—the economics of energy and materials . . . and, second, an economics of information, of entropy, negentropy [exergy], etc. These two do not fit together very well precisely because the units are differently bounded in the two sorts of ecology.

By interpreting the dynamics of information in a system in terms of its bioenergetic flows, Rutledge had melded Bateson's two realms of ecology.

So enthused was I with having discovered how to quantify Atlan's organization that I immediately dropped everything,

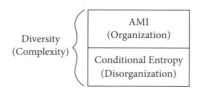

Figure 4.7. The average mutual information and the conditional entropy sum to yield the overall diversity or complexity.

and for the next two weeks did almost nothing else but use a hand calculator to evaluate the magnitudes of the AMI for various network configurations. The calculations convinced me that Atlan had been right. The AMI did indeed correspond to the heuristic notion of system organization. The AMI of a perturbed network invariably was lower than its unperturbed counterpart. Unfortunately, I also discovered that the AMI did not convey the "extensive" nature of the system, i.e., it revealed nothing about the size of the system because the AMI is composed entirely of probabilities, which, in Rutledge's scheme, are estimated by *dimensionless* ratios of flows and sums of flows. In other words, the physical units of the flows all cancel out each other when constructing the ratios. Hence, the final result bears only the dimensions of information (e.g., bits, nats, or Hartleys), and reveals nothing about the magnitude of the physical substrate that instantiates that information.

In practical terms, this meant that, while the AMI of a network representing a microbial community in a Petri dish might be, say, 3.58215 bits per node, the corresponding value for a network representing the biota inhabiting the Serengeti Plains might be 4.08614 bits per node. The difference in the AMIs gives no hint of the fact that these two systems differ in physical scale by several orders of magnitude. It soon became apparent that knowing how well-organized a system is comprises only part of the picture concerning its ability to prevail over time: a well-organized system has an advantage over one that is less structured, but it might still be overwhelmed by another system that is less organized but bigger or more active. Conversely, a vigorous system could be displaced by one that is smaller or less active but better organized. To prevail, a system usually requires a modicum of *both* size and organization. To fully capture the nature of an ascendant system, it becomes necessary to incorporate both size and organization into a single index.

Fortunately, it happened that Tribus and McIrvine (1971)

had recently pointed out this inability of information measures to convey any sense of physical magnitudes. To remedy the situation, they had suggested that investigators using information indices should scale them by some measure of the size of the physical system that incorporates the given information. Although one usually reckons the size of an ecosystem by its aerial extent or its aggregate biomass, Rutledge, Basorre, and Mulholland had used neither of these yardsticks in calculating his probabilities. Rather, they used only system flows and their topological connections to formulate their information measures. To remain consistent with Rutledge and his associates, I, therefore, sought a measure of size that was cast in terms of flows alone.

Placing primary emphasis on flows was not an action unfamiliar to me. Ever since 1979, certain colleagues had been urging me to help promulgate the need to focus on the measurement and use of *flow* variables in the analysis of marine ecosystems (Platt, Mann, and Ulanowicz 1981). Investigators elsewhere had been at work seeking connections between the behaviors of ecological and economic systems. Reckoning the size of a system in terms of its aggregate activity came naturally to them because economists usually give precedence to the activity of an economic community over its stores of capital. In keeping with this attitude, both John Finn (1976) and Bruce Hannon (1973) before him had suggested that the extent of an ecosystem could be gauged by the sum of all the exchanges taking place within it or what they called the "total system throughput" (TST). I decided to follow their lead and scaled (multiplied) the information measure (AMI) by the total system activity (TST) to construct a new index, which I called the *ascendency* of the ecosystem. The new measure was intended to capture in a single index the potential for a system to prevail against any real or hypothetical contending system by virtue of its combined size and organization.

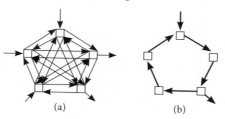

Figure 4.8a. A network of flows having a relatively confused pattern of transfers. *Figure 4.8b.* Another network where the flows of power are more determinate (greater ascendency).

It is perhaps best to think of ascendency as "organized power" because the measure represents power that is flowing within the system toward particular ends, as distinct from power that is dissipated willy-nilly. Almost half a century earlier, Alfred J. Lotka (1922) had suggested that a system's capacity to prevail in evolution was related to its ability to capture useful power. Ascendency can thus be regarded as a refinement of Lotka's supposition that took into account how power was actually being channeled within a system. In Figure 4.8a, for example, power flows in relatively confused fashion among the components (nodes) of the network. In the adjacent Figure 4.8b, by contrast, the flows out of or into any given compartment are more selective and determinate (constrained). In addition, the magnitudes of the flows are greater (as indicated by thicker arrows). The increased power flowing in Figure 4.8b is realized in a more organized way, or, equivalently, the ascendency of network b exceeds that of a.

Beyond just imparting physical dimensions to the ascendency (e.g., $mg\text{-}m^{-2}\text{-}y^{-1}$-bits), our chosen scalar helps further to mitigate Bateson's conundrum: ascendency simultaneously embodies both the economics of material and energy (the magnitude of their activities) in the system as well as the economics of information inherent within the structure of those activities (Collier 1990).[9]

If ascendency should truly gauge the ability of a system to prevail against other configurations, it would then follow that more mature ecosystems should possess progressively greater ascendency. Some years earlier, Eugene Odum (1969) had published a list of twenty-four criteria that he purported characterize more "mature" ecosystems. In going through the list, it soon became apparent to me that almost all of Odum's criteria correlated with increasing ascendency (Ulanowicz 1980). I was eventually motivated to express my phenomenological discovery as follows (Ulanowicz 1997, 75):

> In the absence of major perturbations, ecosystems have a tendency over time to take on configurations of greater ascendency.

This phenomenological statement allows one to quantify explicitly where an ecosystem stands along its developmental trajectory. Furthermore, it bears rough analogy to Lotka's earlier principle of maximal power generation (Odum and Pinkerton 1955). On the other hand, those readers familiar with how optimization criteria are usually stated will be struck by the tentative, almost timorous manner in which the bias toward increasing ascendency was presented. (See also Mueller and Leupelt 1998.) Such ambiguity stands in contrast to the compelling, mechanical-like drive that characterizes the Lotka mandate. Increasing ascendency is explicitly a statistical tendency, and the lack of compulsion owes to the contingent nature of the agencies behind the increase, as was discussed in chapter 3.

For the moment I wish to focus on the fact that increasing ascendency is not the only tendency at work in the dynamics of developing or evolving systems. Apropos, we recall that Rutledge, Basorre, and Mulholland were interested not so much in organization and power as they were in the complement to organization—the "conditional entropy" or the freedom of a network to adapt to novel and unforeseen perturbations. Their particular focus was on how parallel pathways might facilitate buffering in a network (as suggested by Figure 4.6 and by the more numerous pathways in Figure 4.8a over the lack of redundancy in 4.8b).

It happens that parallel pathways are not the only sources of conditional entropy. As the word *entropy* would suggest, thermodynamic dissipations, such as metabolism, also contribute to the measure, as do various inefficiencies in the processing of material and energy or other sources of temporal and spatial incoherence (Ulanowicz 1986, 104ff). I will soon argue that this collection of disorganized phenomena enjoys a rough parity with the organized system activities contributing to the ascendency. I, therefore, chose to scale the conditional entropy by the same TST that I used to impart dimensions to the AMI. The resultant product yielded a measure that is complementary (in the literal, algebraic sense of the word) to the ascendency, and I called the result *overhead*.

Nature in Transaction

A paramount advantage of using information theory to describe any system is that it explicitly parses out the relative magnitudes of order and chaos extant in the system structure. Because ascendency represents the organized power being generated by the system and overhead gauges those activities that are not currently organized but could be entrained into its organization, the sum of these two indices is seen to represent the full capacity for system development. The algebra of information theory shows how this capacity is identical to the diversity of flows after it has been scaled by the TST (Ulanowicz and Norden 1991). After some time, this development capacity tends to approach a limit, becoming almost stationary (Ulanowicz 1986, 105). This limit owes in part to how finely the available sources of resources can be divided (e.g., Pielou 1975). Division cannot continue indefinitely without creating some components that become too small to remain viable. The limit on capacity does not, however, affect how the measure can be partitioned into ascendency and overhead. Either may continue to grow at the expense of the other. That is, a nearly stationary capacity throws ascendency and overhead into unavoidable conflict, so that any increase in one implies a decrease in the other and vice versa. Real systems are the result of an ongoing transaction between the opposing tendencies of both ascendency and overhead to increase.

It has been my experience that this agonism between ascendency and overhead, or equivalently between performance and reliability, has not been widely appreciated outside the field of reliability engineering. Too many persist in thinking that one can have one's cake and eat it, too—that systems can be designed that are both high performance *and* low risk. Yet, examples to the contrary abound: In ecology, those species best able to out-compete (outgrow) their competition invest little energy in avoiding predation, whereas those that develop significant structural and/or biochemical defenses and/or behaviors rarely have the resources left to propagate rapidly. In aviation, beginning pilots are trained on low-performance but stable aircraft. High performance in fighter aircraft can be achieved only through the design of metastable configurations. Too often the value of redundancy is not appreciated until it becomes catastrophically apparent, and, at still other times, the agonism is not recognized even after the fact, as was evident following the 1989 Challenger disaster in the American space program (Ulanowicz 1997, 157). In economics, the competitive edge goes to products that are simply made and hastily assembled, etc., etc. In order better to portray how real systems arise, we must now turn our attention to the nature of the agency behind its increasing order, but we must do so always in the context of a universe that is transactional at its very core.

5

Agency in Evolutionary Systems

Heraclitus was right: We are not things, but flames.
Or a little more prosaically, we are, like all cells, pro-
cesses of metabolism; nets of chemical pathways.
—Karl R. Popper, *A World of Propensities*

The Universal Conversation

The last two chapters have been attempts to answer two complementary questions: "How can things truly change?" and "How
can things persist?" As we discussed at the beginning of chapter 3, these questions bear upon the relationship between change
and stasis—the core issue that separated Hellenistic philosophy into its two major schools. The responses we have provided
are, to an extent, mutually antagonistic. On the one hand, the
second law of thermodynamics, with its rampant stochasticity
(both simple and radical), is constantly disrupting and erod-
ing systems. On the other, cybernetic tendencies, such as autocatalysis, have the capacity to impart order to systems and to
repair disruptions. Thus, the tendencies stand against each other
in dialectical fashion (Ulanowicz 2006).[1] Diogenes relates how
Heraclitus himself saw reality as the result of the dual tendencies
of "building up and tearing down."

Dialectical exchange is dualist by definition (albeit not in the normal Cartesian sense of the word) and conflicts markedly with the desire of most in science to craft a monist description of nature, e.g., formulating a "unified field theory" or attaching a monumental goal to nature—the "variational" approach.[2] It is well known in physics that all reversible laws can be written alternatively as variational principles. For example, Newton's second law of motion is usually presented in the form of a differential equation, but it could as well be expressed in variational format in what is known as Hamilton's principle (Long 1963). Hamilton's principle says that, of all the possible trajectories that an object could take, it will follow the one that minimizes its combination of potential and kinetic energy. All too many ideologies build around such single-track notions. Many economists, for example, push the goal of market efficiency to its monist extreme, making it the *sine qua non* of *all* policy decisions.

That agonistic transactions inhere in living systems provides an initial insight into the gravity of Bateson's (1972) warning against the prevailing approach to problem solving. He deprecated conventional problem solving as driven by a single purpose. That is, normally one formulates the problem, defines the objective, and constructs a method for most directly achieving that goal. The approach sounds eminently sane, and, in those cases where resources are abundant and agencies are sparsely distributed and weakly interacting, it works fine. But when agencies work in close proximity, as they do in complex situations, interferences created by the attempted solution can often cause more overall dislocation than any benefits derived by achieving the goal.

Bateson's aversion to monism spoke clearly to the inability of Newtonian theory to accommodate the full dynamics of living systems, for dialectics cannot be represented in algorithmic (mechanical) fashion (Vanderburg 1990). His concern went beyond merely ignoring indirect effects in ecosystems or disregarding Popper's interferences—the monist approach completely overlooks the fundamental antagonism that both characterizes and supports all life activity. Bateson was acutely aware of the need for a wholly new approach to problem solving. Currently, the approach to mitigating such interference is known as cost-benefit analysis. (Later we will explore another way of mitigating interferences among multiple processes.) No problem solving, however, can ever circumvent the fundamental truth that the dynamics of our complex world are dual in nature. To persist in monist fashion toward a specific goal to the exclusion of countervailing drives leads ultimately to disaster (viz., the Challenger catastrophe).

The probability of arriving at a bad end is all too often inflated by an insistence on using mechanical-like approaches to our goals. After all, if one assumes that we inhabit a world that resembles a clockwork, then there is no inconsistency in bringing mechanical antidotes to bear on problems as they arise. A host of monist ideologies pervade our contemporary world, and all too often their adherents claim support from a mechanical "scientific" approach that ignores the dialectical nature of reality. Growth in the economy is all too often pushed with insufficient regard for the natural subsidies that sustain society. Genetic engineering is almost always pursued for specific ends with virtually no regard for the milieu in which the product must operate and remain viable. Hence, we begin to appreciate how Bateson may not have been paranoid after all; it is indeed feasible to march directly into the jaws of oblivion on the tacit assumptions that support conventional science.

A transactional approach to the dynamics of living systems will not prove easy, however, because more than mere opposition is at work. We have seen, for example, how novel order cannot arise in systems without intervention by aleatoric events. Furthermore, in the last chapter, we saw how autocatalysis can push systems too far toward the mechanical. Some links can grow very strong at the expense of processes that contribute less vigorously to autocatalysis. Those parallel pathways that lose out can wither or vanish, leaving single channels to convey vital flows. Such systems are "brittle" in the words of Crawford Holling (1986), whose path to renown began with the study of budworm outbreaks in spruce forests that had grown too monospecific (Holling 1978). Too large a fraction of the flow of material and energy followed the major pathways associated with the spruce trees. Ascendency grew too much at the expense of overhead. In such a state, any disruption to the dominant, efficient pathways was liable to cause major disruptions. Hence, although the growth and persistence of living systems are driven by structure-building autocatalysis, if efficiency crowds out too much of the remaining stochastic, inefficient, and redundant pathways, the system will respond calamitously to new disturbances.

Conversely, as autocatalysis drives up the levels of aggregate system activity, it concomitantly inflates the rate of system dissipation. This positive correlation between complexity and power dissipation is a well-known problem in computer design (Bose, Albonesi, and Marculescu 2003). In ontogeny, increasing dissipation is strongly correlated with those phases of embryogenesis when the complexity of the embryo is burgeoning (Zotin and Zotin 1997). That more structure in living systems elicits greater entropy production has been discussed by numerous authors (e.g., Ulanowicz and Hannon 1987, Lorenz 2002).

That the larger picture of dialectics goes beyond simple antagonism is an observation attributed largely to Georg Wilhelm Friedrich Hegel. Hegel noted how opposing tendencies can become mutually dependent at some other level of consideration (Salthe 1993). Such dependency at higher levels circumscribes the antagonism between ascendency and overhead. That is, neither can extirpate the other without the whole system going extinct (or radically changing its nature). As Alfred North

Whitehead (1929, 515) so famously put it, "The art of progress is to preserve order amid change, and to preserve change amid order."

If the system performance (order, ascendency) should become too great at the expense of overhead (freedom, reliability), the configuration becomes "brittle" (Holling 1986) and inevitably will collapse due to some arbitrary novel perturbation. Conversely, if the system should become too disorganized (high overhead and little ascendency), it will be displaced by a configuration having greater relative coherence (ascendency) (Ulanowicz 1986, 114ff.).

At the focal level, tendencies are blatantly antagonistic. At another level, however, they can sometimes require and even support each other. One cannot simply push one side of the transaction to the exclusion of the other. As mentioned, many economists pursue the goal of market efficiency to its monist extreme. In the process, they ignore that which imparts reliability to a community, such as functional diversity and equity of wealth (e.g., Daly 2004). With their zeal, they unintentionally set society up for a fall. If the reader takes away only one idea from this whole thesis, it should be that pursuing a single (variational) goal, while failing to consider its agonistic counterparts leads invariably to a bad end. Directions are essential elements of the evolutionary drama, but, like the propensities that give rise to them, they *never* occur in isolation.

The Organic System

It happens, then, that causality in complex systems is more complicated than what has been allowed under the strictures of closure. We have just discussed why causality in biotic systems cannot be interpreted entirely as rigid, mechanical, and deterministic. This is one reason that I prefer the term *top-down influence* over the more rigid notion of top-down causality.

Rigid causality anywhere begins to look more like the exception than the rule. But there is another reason behind my reticence—I wish to avoid focusing too much on the organism. I have already mentioned reasons that I think ecology is a richer theater than ontogeny in which to pursue a deeper understanding of the growth and development of living systems. I worry that our fixation upon the organism as the primary token of a living system has needlessly complicated our discourse. In particular, the mechanical-like constraints that characterize ontogeny tend to draw one's attention away from the dynamics that are at the core of organic behavior (Ulanowicz 2001). As we saw when ecologists rejected Clements, the rigid nature of constraints in organisms understandably caused many to eschew the organism as a metaphor for ecological, economic, and social systems—an analogy known as *organicism*. Even the later approach to living systems known as autopoesis (which bears many parallels to process ecology) started with its focus upon the organism and adopted an explicitly mechanical standpoint (Maturana and Varela 1980). Top-down influence in organisms, however, is far stronger than that encountered in ecosystems or, for that matter, in social systems. The strength of control in organisms has enticed any number of would-be or actual dictators into using the organism as an analogy for their government, in an attempt to justify their rigid control of the "organs" of society (e.g., press, labor, etc). This practice has unfortunately given organicism a bad name.

Clements, however, was not entirely wrong in suggesting parallels between ecosystems and organisms (Ho and Ulanowicz 2006). It is just that the latter is characterized by far too many rigid constraints (viz., Ho 1993). In order to facilitate legitimate analogies without invoking the specter of rigid top-down control, I have suggested that those ensemble living systems that exhibit organic-like behaviors, but are more flexible than organisms, be referred to as *organic systems* (Ulanowicz 2001). That is, organic systems exhibit some degree of top-down selection and system coherence, but such influence is less strict and programmatic than what one encounters in organisms, where development of the system follows an inflexible script that usually includes the construction of an integument to surround itself.

The Organic Fold

It appears, then, that top-down influence is a defining characteristic of higher-level systems, but reductionism appears to work well among the lower, less complicated levels of the physical realm. The reason reductionism remains appropriate at lower levels is that the assumption that the smallest particles of decomposition are both simplest and longest lasting is a good approximation of reality there. As one proceeds from the simplest subatomic entities, such as quarks, to the larger scales of atoms and molecules, combinations of particles become both more complex and endure for progressively shorter lifetimes. Both trends render the compound entities amenable to treatment under the Newtonian assumptions.

Now, we usually think of higher-level, ensemble systems as more complex than those below them because the perspective we adopt is to look vertically down through the hierarchy, so that the complexities at the various levels become additive. But, first, we should be cautious because not all complexity appears in organized form. Second, the ensemble systems appear quite differently when one regards them from a horizontal standpoint, i.e., at their own focal level. From this angle, much, if not most, of what one sees are copious degrees of freedom, and the organization at that level does not appear as developed or rigid as it does at the next lower level, which is the organism itself. In addition, as we noted in the last chapter, whatever organization as may exist at the ensemble level tends to endure longer than do its constituents. In other words, the highest-level entities are both simpler in their organization and longer lasting than the ones immediately below them—the exact inverse of circumstances among lower physical levels. This inversion suggests strongly to us that explanations at macroscales should be easier to formulate in top-down fashion.

In mathematics, when one encounters a reversal of directions (or more accurately of gradients) at two separate points along an axis, one of two things must happen between those points: either the property of interest takes on an extreme value (a maximum or minimum), or it undergoes a catastrophic discontinuity (a discontinuous jump or pathological singularity, either of which serves as a barrier that degrades prediction and control from the opposite side).[3] For reasons that should become clear in the next section, I think the catastrophe is more likely. It is well known that most systems behave quite differently on opposite sides of a catastrophe. Here the image of a phase change from solid to liquid, a simple type of physical catastrophe, comes immediately to mind. So as to avoid any confusion between the common and technical meanings of *catastrophe*, I have elected to call the transition the *organic fold*. (One of the basic discontinuities in catastrophe theory is called a *fold*.) Hence, I am suggesting that agencies in natural systems differ radically in their dynamics, on opposing sides of the organic fold.

The reader could well ask where along the axis of system complexity the organic fold might lie. For the time being, one can only guess at its location, and my conjecture is that dynamical behavior changes radically as soon as the system begins to contain polymers more than 30–50 monomers in length and of sufficient variety.[4] My reasoning is that, depending on the variety of distinguishable monomers, the point at which the combinatorics of the system begin to overwhelm the capacity of laws to determine future events should lie somewhere in this range of molecular complexity. Beyond this divide, one would expect that processes supplant laws as the most effective determinants of system structure and behavior. As we have seen, the likelihood that singular events and emergent behaviors will occur grows markedly once the number of possible combinations of entities far exceeds the number of existing material units. A molecule of this complexity should be of sufficient diversity that the substitution of one or a few monomers will not *necessarily* induce a radical change in its form or in the dynamics in which it participates. (Our fixation with large macroscopic changes that are induced by the substitution of single monomers in nucleotides or proteins focuses on exceptions, not the rule.) At the same time, a molecule of 30 or 50 monomers could be sufficiently complex as to be unique. (In reckoning the molecules requisite for life on earth, one need not consider the full 10^{81} simple particles in the entire universe, as Elsasser did, before one encounters uniqueness. A figure closer to the number of atoms in the incipient earthly biosphere should suffice.)

Evolution Is Like a Muscadine Grapevine

Our focus in this essay is on systems that lie above the organic fold in the realm where top-down influence is most likely. That a domain might exist in which top-down formal influence is the predominant form of causality lends a very different hue to the nature of evolution than one finds in contemporary theory (Ulanowicz 2004a). The renowned philosopher Daniel Dennett (1995), for example, described the evolution of ever more complex biological entities as analogous to "cranes built upon cranes," whereby new features are hoisted on to the top of a tower of cranes where they become available to build yet another crane that repeats the process. Dennett cautioned his reader not to consider any influence from above that cannot be connected with smaller-scale agencies that lie below. This prohibited a type of agency he calls a *skyhook*. Dennett's concern about skyhooks is understandable because they are too suggestive of transcendental influence. Overall, however, I eschew his metaphor as misleading. Like Elsasser and Bateson, I believe that many, if not most, mechanical analogies are procrustean, minimalist distortions of the dynamics of living systems. Paraphrasing the analogy of Oliver Penrose (2005), living dynamics only look rosy (mechanical) because most insist on looking at them through rosy (Newtonian) glasses. Dennett's simile diverts one's attention from their true nature.

Dissatisfied with Dennett's analogy, I felt the need to find a more organic simile for the evolutionary process. I was lost in thought on the subject while puttering distractedly in my garden, when I experienced the only "Eureka!" phenomenon in my life (Ulanowicz, 2004a). While pulling weeds, I was suddenly was drawn to a muscadine grapevine that has grown on the corner of my garden fence for the last twenty-five or so years. In the initial years after I had planted it, the lead vine had climbed the fence to become a central trunk that fed a lateral growth of grape-bearing vines (Fig. 5.1a). Eventually, the horizontal vines had let *down* adventitious roots that met the ground somewhere less than a meter from the trunk (Fig. 5.1b). Then, in the last few years, the main trunk had died and rotted

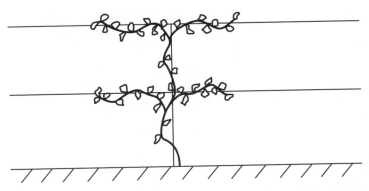

Figure 5.1a. Young muscadine grapevine with central stem and branches.

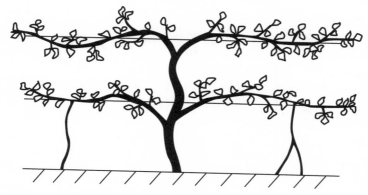

Figure 5.1b. Grapevine several years later, having developed adventitious roots to the sides of the main trunk.

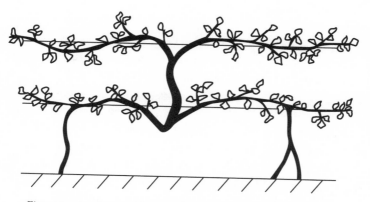

Figure 5.1c. Same grapevine two decades later. Original trunk has rotted away, but vines are sustained by adventitious root system.

away completely, so that the framework of vines was being sustained by the newer connections, which themselves had grown to considerable thickness (Fig. 5.1c).

It struck me like lightning that here was a more appropriate metaphor for the dynamics of evolution! The muscadine plant represents an evolving system across several hierarchical levels.

No skyhooks are involved because the system always remains in contact with a foundation of bottom-up causalities that remain necessary to the narrative. Consistent with the scenario of process ecology, however, later, higher structures create new connections that eventually replace and/or *displace* their earlier counterparts. Such displacement is a key element in solving the enigma of how emergent structures and dynamics can influence matters at lower levels that had played a role in their own appearance.

Such top-down influence commonly has been called *supervenience*, but that term has been overused to the point of confusion.[5] We shall refer to the phenomenon instead as *suprafacience* and note that it has been exceedingly difficult to specify how such influence might be possible (Peterson, forthcoming). The answer to this conundrum, as prompted by the simile of the muscadine grapevine, is surprisingly simple—higher-level agencies don't simply affect their progenitors: they replace them! To see how, it becomes necessary to pay close attention to the temporal sequence of events.

As the first step in the emergence of a new dynamical entity, the upper-level dynamics (such as autocatalytic configurations) are first brought into existence by whatever processes, laws, or chance events as happen to be available (the initial vine and branches). Once the autocatalytic loop has been engaged, however, it begins to exercise its own agency in any of the several ways that were discussed in the last chapter. It is already no longer acting completely at the behest of the sources that brought it into being (partial autonomy). Chance new connections to resources may appear (the adventitious roots). Autocatalytic feedback will reward any of the new connections that provide greater sustenance and/or persistence to the upper structure (the thickening of the new roots). Eventually, some of the new connections will become far more efficient than the original ties that brought the upper structure into existence, so that the new will replace the old (the disappearance of the initial trunk). The full sequence of such displacement I call *temporal suprafacience*.

We have already given one putative example of temporal suprafacience: the appearance of DNA and its eventual displacement of all earlier methods of information storage (Deacon 2006). Other examples include the activation (via acetylation) and inhibition (methylation) of nucleotides during ontogeny that preserves regulatory states even after the initial activators/repressors are no longer present (Davidson 2006).[6]

All metaphors are, by definition, imperfect and should never be taken too literally. For example, some readers might be quick to point out that the habit by muscadine grapes of putting down adventitious roots has been encoded in the genome of the species. This is true, but it lies entirely beside the point being made about dynamics. Furthermore, to give Dennett's simile due credit, his metaphor does emphasize the role that enduring history plays in evolution. My analogy, by contrast, highlights how active agency can displace some (but never all) of a system's history. Reality is likely a combination of both scenarios. Similarly, caution must be exercised when focusing upon downward influence that one not lose sight of the fact that causality in nature is necessarily a two-way conversation between top-down and bottom-up agencies.

Suffice it as my last word on the subject that a more organic metaphor is needed to represent the tradeoff between history and novelty in evolution.

The Preeminence of Organic Agency

Declaring that configurations of processes sometimes act in suprafacient ways is a drastic—many would say, heretical—statement, but it absolutely central to our understanding of what it means to be alive. To underscore the significance of action by a pattern of processes, I cite the example of the deer that has just been shot by a hunter. Enzo Tiezzi (2006) asks what is missing in the dead deer that had been present in the minutes before it was shot. Its mass, form, bound energy, genomes—even its molecular configurations—all remain virtually unchanged immediately after death. What had ceased with death and is no longer present is the configuration of processes that has been coextensive with the animated deer—the very agency by which the deer was recognized as being alive.

Quantifying Growth and Development

New ways of envisioning causalities in nature are helpful for understanding phenomena, but if one wishes to engage fully in the scientific enterprise, it becomes necessary at least to describe how one might translate prose into mathematical statements. As regards this task, I should count myself fortunate because, as I recounted in the first chapter, my entire project on process ecology began with a phenomenological (i.e., quantitative) discovery—namely, that most of the attributes that Eugene Odum had identified as characteristics of well-developed ecosystems could be encapsulated into a single quantitative index—the network ascendency (Ulanowicz 1980, 1986). That phenomenological observation led me to focus upon configurations of processes as the crux of agency in living systems, and a full verbal narrative ensued. It is now necessary to return to the ascendency to locate exactly how that measure expresses the constraints engendered by autocatalytic agency.

To summarize some of what we have discussed about development, the qualitative changes that autocatalysis brings about in a system network are twofold. First, autocatalytic feedback, by definition, works to increase the overall level of system activity. Second, it tends to increase the magnitudes of those connections that are most active in any particular autocatalytic process at the expense of nonparticipating processes and rela-

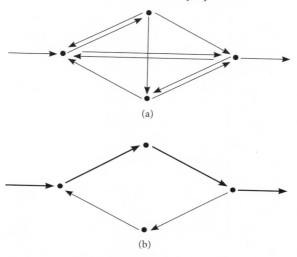

Figure 5.2. A schematic of the combined extensive and intensive effects of autocatalysis within a hypothetical 4-component system. (a) An inchoate network of flows. (b) The same network once total activity has increased (represented by thicker lines on the arrows) and after inefficient and redundant flows have been pared from the network.

tionships (Figure 5.2). The first effect is upon the "size" of the system and would be measured by what in thermodynamics is called an *extensive* variable. The second effect deals only with the relative sizes of the transfers and is thereby qualitative or independent of the system size. It would be tracked by what is called an *intensive* variable in the parlance of thermodynamics.

We emphasize as how the intensive effects of autocatalysis can be adequately gauged only when one has information on the relative magnitudes of system transfers. This requirement is not met by most of what one encounters in today's literature on network theory. Most contributions still deal with equiponderant, undirected connections between the elements (nodes) of any system, or, at best, some indication of the direction of an

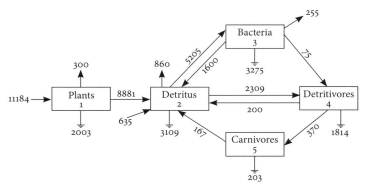

Figure 5.3. Schematic of the total suite of energy flows (kcal m^{-2} y^{-1}) occurring in the Cone Spring ecosystem. Arrows not originating from a box represent inputs from outside the system. Arrows not terminating in a compartment represent exports of useable energy out of the system. Ground symbols represent dissipations.

effect is given. I reiterate how the tradition in ecology from the very beginning has been to distinguish among the magnitudes of the various flows. Not knowing the transfer magnitudes, one is unable fully to distinguish and to quantify the changes elicited by autocatalysis as depicted in Figure 5.2 (more on this in the next chapter).

Thus, although it is of some help to know that predator j is consuming prey i, one seeks in addition to estimate how much is being transferred from prey i to predator j per unit time. Let us call that amount T_{ij}. A simple example of a quantified ecological network is illustrated in Figure 5.3, which presents the estimates of Tilly (1968) for the magnitudes of all of the transfers taking place among the five principal components of the ecosystem inhabiting a small freshwater spring in Arkansas called Cone Spring. We see, for example, that component 4 consists of "detritivores" or those invertebrate species that eat dead organic material (the "Detritus" of component 2) and its attached bacteria (component 3). Detritivores ingest 2309 kcal

$m^{-2} y^{-1}$ of the former and 75 of the latter, so that $T_{24} = 2309$ and $T_{34} = 75$ kcal $m^{-2} y^{-1}$.

As mentioned in chapter 1, the total activity occurring in the Cone Spring ecosystem is reckoned simply by adding up the magnitudes of all the flows, yielding an extensive variable called the total system throughput, or TST. In the case of Cone Spring, all the flow magnitudes add up to 42,445 kcal $m^{-2} y^{-1}$. Presumably, any stimulus to the activity of the Cone Spring ecosystem resulting from autocatalysis would be reflected by an increase in this sum.

The intensive action of autocatalysis that resembles "pruning" (Figure 5.2b) is somewhat more difficult to quantify. As related in the first chapter, both the organized and chaotic aspects of the pattern of connections are quantified using information theory (Rutledge, Basorre, and Mulholland 1976). It is not the purpose of this narrative to examine the intricacies of information theory here. The reader interested in the mathematical details is referred to Hirata and Ulanowicz (1984), Brooks and Wiley (1988), Ulanowicz and Norden (1990), and Ulanowicz (2004b). All the pertinent information measures are computed by combining the logarithms of the various fractional flows into and out of each compartment. To give the reader a rough feel for these measures, I offer the following three patterns of how three flows might exit a hypothetical compartment (those readers uninterested in the calculation of AMI may skip the next five paragraphs):

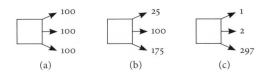

Figure 5.4. Three distributions of three hundred total units of outflow from a hypothetical compartment.

In configuration (a), one-third of the total flow exits via each route. Taking the negative of the logarithm (base 2) of the first fraction (1/3) yields what in information theory is called the *surprisal*, which, in this case, works out to 1.585 bits. (A *bit* is the amount of information required to resolve a single binary decision.) The calculation is referred to as surprisal because, when the probability of an event is very small (near zero), the negative of its logarithm becomes very large, i.e., one is very surprised whenever the rare event occurs. Multiplying the surprisal by the probability itself (1/3) yields the measure of indeterminacy (0.5283 bits). The measure of indeterminacy is largest for probabilities in the middle range because outcomes with probabilities nearer either extreme (1 or 0) are presumed to be under strong constraints keeping them near those extremes. (They either approximate determinate situations [as in Figure 2.3] so that their values work out to be near zero, or they are being kept minimal due to vigorous external influences.) When the same two calculations are applied to the other two flows in (a) and all three results are summed, one arrives at a value of 1.585 bits representing the average indeterminacy of the three flows. We note that, if we raise the logarithmic base (2) to the power 1.585, the result is exactly 3.0, confirming that, indeed, three equiponderant flows are leaving the compartment.

When one calculates the overall indeterminacy for configuration (b), however, a lower value of 1.281 bits results. This infers that situation (b) is less indeterminate (more constrained) than (a), and it is plausible that some (possibly unknown) constraints are at work causing more medium to leave via the bottom route than exits via the top two. The decrease in indeterminacy (1.585–1.281 = 0.304 bits) can thus be regarded as quantifying the intensities of all the constraints (both overt and hidden) that are skewing flow toward the bottom exit. This difference is also a measure of the information embodied in the system.[7]

Taking the base 2 to the power 1.281 gives a value of 1.641 as the measure of the "effective number of flows" out of the box. Topologically (qualitatively) speaking, three flows can be identified as exiting the compartment. More than half of the flow exits via the bottom arrow, however, and, in our calculations, it gets weighted more than the others. Because egress is dominated by the bottom route, any quantum of medium effectively sees fewer than two ways out of the compartment. The flows of carbon out of the sea nettle component in the Chesapeake ecosystem (Baird and Ulanowicz 1989) resemble this intermediate example. Of the total of 1600 mgCarbon/m²/y passing through this species, 188 units fall as particulate carbon to the bottom, 447 remain as particles suspended in the water column, and the bulk, 1,076 units, are respired as CO_2 dissolved in the water column. The effective number of pathways out of the compartment is 1.951, and the constraint that skews the value below 3 is the relatively high respiratory demand by the nettle.

The effect of weighting becomes even more apparent in configuration (c). There one sees that very strong constraints are at work (such as might arise from autocatalysis) to channel most of the flow to leave via the bottom pathway. The measure for indeterminacy has fallen in this case to a scant 0.089 bits, and, as a consequence, the intensity of constraint has risen to 1.496 bits. The effective number of flows turns out to be 1.064, signifying what is obvious by looking at the diagram—that there effectively is little more than one way out of the compartment. One would expect the outputs from top predators to resemble this pattern in that almost all of their production leaves the system as respiration and very little flows to other consumers.

The calculation of the average mutual information (AMI) is very similar to how we calculated the intensities of the constraints extant in each compartment of Figure 5.4, only the averaging procedure is extended to the entire system (Hirata and Ulanowicz 1984). For the Cone Spring example in Figure 5.3, the AMI turns out to be 1.336 bits. This value is scaled by the total system throughput (42,445 kcal m^{-2} y^{-1}) to yield a magnitude of 56,725 kcal-bits m^{-2} y^{-1} for the ascendency of the Cone Spring network.

To give the reader some appreciation for how autocatalytic action is usually manifested as increasing ascendency, three hypothetical stages of development are provided in Figure 5.5. All three networks exhibit the same total system throughput (120 arbitrary flow units). In configuration (a), much of the activity appears as flow into and out of the system. Internal circulations are small, and guiding constraints are weak (equiponderant amounts flow from each compartment to all the others). The ascendency value is forty-three flow-bits. In network (b), a smaller proportion of flow consists of exchanges with the surroundings, and the transfers to other compartments are somewhat less ambiguous. (Flows from each compartment connect with only two of the remaining three compartments.) The ascendency (for the same amount of total activity) has doubled to eighty-six flow-bits.

The differences between configurations (a) and (b) illustrate how more medium is progressively being cycled within the system (rather than passing through it) along ever more distinct pathways. If these tendencies were allowed to reach their theoretical maximum, the result would look like that shown in Figure 5.5c. In this hypothetical limit, no exchange takes place with the external world, all flows become equal, and only one flow enters and leaves each compartment. If represented as a matrix, this endpoint would resemble Table 3.2, and the shift from (b) to (c) in Figure 5.5 would be analogous to the transition from Table 3.1 to 3.2. We note for later reference that the hypothetical endpoint represents a system in full thermodynamic equilibrium—it is totally isolated. Nothing flows into or out of it.[8] Furthermore, the internal relationships have become stiff and mechanical.

We know, however, that no biotic entity is in thermodynamic equilibrium (only dead ones). All living systems must

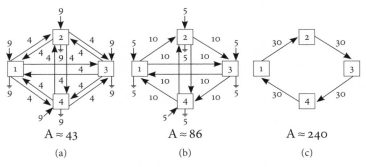

Figure 5.5. The progression of a hypothetical simple system:
(a) a very inchoate configuration, (b) a more constrained
one, and (c) the maximally constrained situation.

import high-quality energy (nutrition) from their environment
and release lower-quality energy (wastes) to their surroundings. These same requirements preclude living networks from
ever reaching a configuration of maximal ascendency (Figure
5.5c). As defined earlier, the degree to which they fall short of
their maxima is called *system overhead*. A major contribution
to the system overhead is the number of parallel or redundant
pathways within the system, and we note how indeterminacy
progressively decreases as one goes from the configuration
depicted in Figure 5.5a to that in 5.5c (see also Figure 4.6). Other
contributions to overhead include inefficiencies at processing
medium, incoherencies in temporal behavior, and stochastici-
ties in spatial distributions. In short, all those characteristics
that do not contribute to organization (ascendency) add to the
overhead. The calculated overheads for the three networks in
Figure 5.5 are 462, 384, and 0 flow bits, respectively. (As ascendency increases, overhead declines.)

So, ascendency tells one how well the system is processing
medium, i.e., how well it is performing or how well organized is
the power flowing within the network. Overhead, on the other
hand, encompasses all the disorganized elements and behaviors

that remain. It should immediately be noted, however, that such incoherent behaviors also represent degrees of freedom and flexibility that could be used by the system in the event of novel disturbance to replace functions in the previously unperturbed configuration. Thus, although overhead represents inefficiency and incoherency under normal operation, it also serves as strength-in-reserve to ensure system reliability.

As noted in the first chapter, the sum of a system's ascendency and its overhead is called its *development capacity*. There we argued as well that variation in development capacity inevitably diminishes in relation to those of either ascendency or overhead. A relatively constant capacity means that performance (ascendency) and reliability (overhead) become mutually exclusive—a rise in one will cause a drop in its complement (as we first saw with the networks in Figure 5.5). Hence, the values for capacity, ascendency, and overhead allow one to quantify exactly where the system stands along its trajectory of growth and development.

Locating Vulnerabilities

Not only does ascendency provide a measure of how well the system is performing, but the individual contributions of the compartments and exchanges to the overall ascendency also reflect the importance of each element in the overall system dynamics. Serendipitously, it is extremely easy to calculate how sensitive the total system ascendency is to changes in compartmental stocks or individual flows (Ulanowicz and Baird 1999, Ulanowicz 2004b). Whenever the system ascendency exhibits significant sensitivity to change in a particular stock or flow, that element is likely to be a bottleneck or limiting element to system dynamics.

Ulanowicz and Baird (1999), for example, used the sensitivities of the ascendency to changes in stocks and flows to pinpoint the factors that control nutrient dynamics in the Chesapeake mesohaline ecosystem. Among other things, they discovered that different compartments were controlled by different nutrients. For example, it has long been known that primary producers are limited by nitrogen during the summer season (D'Elia 1988)—a fact that was confirmed by the ascendency sensitivity analysis. Surprisingly, however, the sensitivity calculations revealed that bacteria and fishes were most limited by phosphorus during the same season. (This makes sense in light of the need by these organisms for phosphorus to create ATP and bones, respectively.) The same sensitivity analysis can be applied to heterogeneous systems (either living or nonliving) to identify the most critical points of a system in space and time (Ulanowicz and Baird 1999, Kikawada 1998, Ulanowicz and Zickel 2005).

In this fashion, sensitivity analysis of ascendency identifies the species, times, and places where a system is most likely to be vulnerable to disturbance. It only rarely happens that a system collapses or radically changes its dynamics in response to a point-wise perturbation (even when that point is among its most sensitive). It can be another matter entirely, however, when several points of vulnerability are matched by a complex disturbance in lock-and-key fashion. The system might then either collapse or be thrown into some *emergent* mode of behavior. That is, systems are likely to continue in a reasonably stable mode of behavior for a long while, only to change abruptly in response to a rare or possibly unique configuration of events. These dynamics have been described by evolutionary theorists Gould and Eldredge (1977) as *punctuated equilibria*. Such transformations are totally natural, albeit rare occurrences in the dynamics of process ecology.

Not all transition states represent emergent events, and some are reasonably common in nature. For example, the transitions from terrestrial forest to salt marsh to estuary occur in punctuated fashion as sea level rises (Brinson, Christian, and Blum 1995). Along this geographic cline, one can identify five stable plant communities, each of which is reasonably resilient to gradual rise in sea level. Transition from one community to another is usually in response to some change more abrupt than sea-level rise, such as brackish water intrusion, tidal creek encroachment, or corrosive currents and waves.

Finally, it is worth mentioning that some readers may have noticed a strong resemblance between the figures I have presented in these last sections and those one encounters in the literature on ecological simulation. I need to stress, in case it is not already apparent, that such resemblance can be misleading. Network analysis and simulation modeling do, indeed, begin from the same database. They both require the identification of the relevant compartments, along with some estimate of their densities and the topology of weighted exchanges among them. Model simulation then proceeds to construct a collection of mathematical statements describing how the changes in each compartment depend upon other compartments and exogenous factors in order to forecast future densities and exchanges. By contrast, network analysis attempts no *a priori* assumptions concerning how changes take place. Rather, the assembled network is analyzed to assess how much each component is contributing to overall system organization and persistence and proceeds to identify the controlling exchanges. Whereas simulation modeling is analogous to studying the system's *assumed* "physiology," network analysis quantitatively articulates its *demonstrated* "anatomy."

6

An Ecological Metaphysic

Let us define ecology as the study of the Universe.
—G. Evelyn Hutchinson, quoted in "A Global Vision"

Axioms for Life

The preceding chapters have ranged over a welter of unconventional topics. It might help at this point if we were to pause and take stock of the emerging form that process ecology is taking. At its very core lie three fundamental postulates:

I. The operation of any system is vulnerable to disruption by chance events.

II. A process, via mediation by other processes, may be capable of influencing itself.

III. Systems differ from one another according to their history, some of which is recorded in their material configurations.

As I have intimated earlier, none of these statements in itself is even remotely radical, unnatural, or contradictory to common sense. It is only in comparison with what historically has anchored the scientific enterprise that they could be regarded as problematical. For example, closer inspection reveals that each of the three postulates negates one of the original Newtonian axioms. The first is a direct contradiction of the Newtonian assumption of determinism. It is unlikely to be embraced by those whose primary interest in science is, following Bacon, to control nature. The second postulate becomes controversial as soon as one realizes that it dispenses with obligate reductionism, which as we have seen is supported by closure and atomism. Implicit in the statement is a renunciation of atomism—that all explanation is to be cast in terms of identifiable *sub*units. At least some causality can arise *at* the level of observation. We discovered later that the effects of feedback can be impressed upon lower levels, which violates the spirit, if not the letter, of closure. The third axiom is perhaps oldest and most familiar, as science has been confronted with history since before the time of Darwin (Buffon 1778, Cuvier 1825). But history extirpates reversibility. Finally, the earlier faith in universality is, like our previous confidence in atomism, now absent from our formulation.

Corollary Shifts

Acceptance of these primary assumptions initiates several other important shifts in emphasis. The acknowledgment of chance as a reality and not just an illusion leads one to question whether laws can provide the full story of how living systems arise and are sustained. It's not that laws are ever violated. No account of evolution will ever defy the laws of gravity or columbic attraction, and it assuredly remains important how a system parses the energy within and flowing through it (e.g., Jørgensen and Mejer 1979, Schneider and Kay 1994). Laws retain their power to constrain and guide. It's just that we recognize that laws can be overwhelmed by the variety inherent in both chance and existing order. Combinatorics overwhelms the particulate: there is such a multiplicity of heterogeneous configurations that can satisfy the same lawful constraints that laws forfeit their power to specify precisely which of the manifold alternatives will prevail. That specification now lies in the domain of process or, to put a finer point on it, with configurations of processes. Processes exert the power to return nonrandom response to random inputs. We are thus led to recommend:

1. In order to understand living systems, emphasis should shift away from fixed laws and toward the description of process.

Laws deal with efficient causes that take the guise of forces, and the Newtonian prescription always couples those forces with matter or objects. Just as complex chance and explosive heterogeneity erode the power of laws to specify, they likewise diminish the efficacies of forces, mechanisms, and objects as explanatory tools in biology. We were able to rescue explanation from the eclipse of law by invoking processes. In similar manner, Popper has suggested that the notion of force be expanded to the more general concept of propensity. Furthermore, Popper (1990) was adamant that propensities never be considered in atomistic fashion. Propensities never occur alone, he insisted, but always in relationship with other processes—as configurations of propensities, so to speak. We are thus led to recommend another major shift in emphasis:

2. Relevant agencies in living systems reside more with configurations of propensities than with explicit physical forces or their attendant objects.

This redirection of emphasis occasions a major reorientation in our thinking. Suddenly, configurations of processes or propensities rather than objects become the focus of our attention in explaining how and why things happen in biology. This shift takes a little getting used to, accustomed as we are to dwelling upon objects as the crux of all change, e.g., genes *direct* the development of organisms. Certainly, organisms (as they exist today) require genes to develop. But, as we have suggested, these material elements play relatively passive roles in comparison to the activities of the networks of enzymatic processes that actually interpret the encoded information and "incorporate" it.

Yet one other important change in thinking is demanded by process ecology: recognition that development is the outcome of dual and opposing tendencies. Here we plainly are not referring to Cartesian dualism, so disdained by contemporary thinkers, but rather that stressed by Heraclitus and Hegel:

3. Patterns and forms in the living realm result from transactions between agonistic tendencies. Processes that build organized activities are continually being eroded by dissipative losses. While these tendencies oppose one another in the near field, they are seen to be mutually obligatory under a wider vision.

The important point here is never to push single goals too far because doing so invariably leads to system catastrophe.

If the reader should have qualms about adopting the process viewpoint, he or she should be comforted to learn that some old conundrums simply evaporate when viewed through the new lens. Striving, for example (when it was mentioned at all under the old conventions), mysteriously arose from unspecified associations of genes. But how do the material loci actually induce the activity? Under the ecological metaphysic, centripetality derives from self-influence, and the imperative to strive and compete emerges out of centripetality. Rather than a mystery or anomaly, striving is central to the view seen through the window of process ecology. The new perspective orders attributes in a new priority as well. Mutuality, for example, is seen to be ontologically prior to competition. Finally, the phenomenon of emergence flows smoothly out of the new dynamic. Suspicion no longer enshrouds it—it forms a perfectly natural element of the dynamical landscape (Ulanowicz 2007).

Chance as Actor

With this tapestry of process ecology now spread out before us, we return to consider each element in its turn. Beginning with chance, science has largely relegated its action to the microscales, where all occasions of the aleatoric are presumed to act simply and independently. This uncomplicated view allows an investigator to render the effects of chance intelligible via the application of statistical methods, thereby appearing to restore a degree of control overall. We are all familiar with situations, as diverse as the properties of gases or life-insurance survival tables, where probability theory does indeed serve us well: using statistical mechanics, we can estimate the viscosity of neon gas at any temperature and pressure; using survival tables, we can set up a successful insurance business. Elsasser, however, has warned us against the hubris of thinking all is now copacetic by questioning those simplistic assumptions about chance and by making us aware of ubiquitous complex chance events that elude treatment by statistics. We can forget about any equivalent of survival tables that will tell us when a new species will emerge from an existing one or what the exact nature of the novel successor will be.

In chapter 3, we discussed two reasons that chance should be considered real (ontic) rather than illusory (epistemic), namely, (1) heterogeneity and (2) combinatorics. Countervailing arguments favoring chance as epistemic usually cite examples in which participating elements are both simple and identical—a veritable Lucretian universe. The interaction of identical elements accords with the notion of lawfulness because laws are built upon the continuum assumption, which, in turn, is logically congruent with operations between homogeneous sets (Whitehead and Russell 1913). As soon as the interacting elements become heterogeneous, however, the logical basis for lawfulness disappears.

Furthermore, the applicability of lawfulness becomes even more doubtful whenever the participating heterogeneous elements are themselves complex. In chapter 5, under the section "Evolution Is Like a Muscadine Grapevine," for example, we argued how configurations of processes are not wholly the consequences of their primitive constituents because upper-level configurations can exert suprafacience upon their lower-level components. It follows that any notion whereby collections of complex configurations (as appear in ecology, sociology, or economics) are driven entirely by their smallest elements according to laws (e.g., as in sociobiology) severely taxes credibility.

Then there is the specter of combinatorics. Elsasser tacitly invoked Laplace's formulation of probability to demonstrate that there exist far, far more potential combinations of distinguishable events than the known universe has material or time to accommodate. For example, there exist approximately 10^{870} combinations of four hundred distinct elements, which mean the Laplacian probability of any particular combination is 10^{-870}. With at the very most only 10^{106} physical representations possible, it follows that an astronomically overwhelming majority of the potential configurations cannot occur. Most categories remain perforce empty, and only rare configurations are populated by a single token. The fraction of categories with two or more configurations is vanishingly small. Not only does this rarified distribution defeat the frequentist approach to defining probability, but it also renders nonsensical the notion of the continuum—a necessary precondition for law.

It is necessary to recall that the tenet of causal closure was formulated as philosophical dogma before the inflationary nature of the universe had been discovered or even before Carnot's observations on irreversible phenomena.[1] Despite these later revelations, adherence to closure persists—witness David Bohm's (1989) assertion of implicate (hidden) order over and against the "Copenhagen interpretation" (reification) of quantum probabilities (Heisenberg 1927, Bohr 1928). Furthermore, there are some who see in chaos theory a vindication of determinism: the equations that give rise to chaos are indeed deterministic, but they generate phenomena that *appear* chaotic (Ulanowicz

1997). Perhaps, the proponents of implicate determinism would suggest, chaos is emblematic of all aleatoric phenomena. Chance only appears indeterministic, but it derives from precise laws. It is worth noting that such arguments reasserting determinism via chaos theory always follow time forward. Looked at retrospectively, however, chaos theory also implies that it is impossible to trace antecedents back beyond a certain duration: that is, if a given phenomenon had been initiated by some small event in the past, that cause can never be precisely identified (Ulanowicz 1997). In either temporal direction, the limitations on predictive or postdictive determination via chaos theory render impossible any definitive test of determinism.

Then there is the second law of thermodynamics, which caused so much trouble for the physics of the nineteenth century. It is hardly ever pointed out that the irreversibility demanded by second law bespeaks a degree of causal openness. This conclusion is nowhere more apparent than in the version of the law proposed by Constantin Carathéodory (1909), mentioned earlier. By saying that arbitrarily close to any equilibrium state lays another state that cannot be reached via a reversible pathway, Carathéodory was implying that phase space is discontinuous at the microscale—and such distributed discontinuity is inimical to closure.

For me, the most compelling case against causal closure is by analogy to logical closure. As most readers are probably aware, Kurt Gödel (1931), working with number theory, demonstrated how any formal, self-consistent, recursive axiomatic system cannot encompass some true propositions. In other words, some truths will always remain outside the ken of the formal system. The suggestion by analogy is that the known system of physical laws is incapable of encompassing all real events. Some events perforce remain outside the realm of law, and we label them *chance*. Of course, analogy is not proof, but I regard Gödel's treatise on logic to be so congruent with how we reason about nature that I find it hard to envision how our construct of physical laws can possibly escape the same judgment that Gödel pronounced upon number theory.

As for the scale at which ontic chance can happen, the key word here is *ubiquitous*. We discussed in chapter 3 how nothing stands in the way of its appearance at macroscopic levels, where complexity abounds. In fact, complex chance is even *more likely* among the complexity of macroscopic biotic phenomena, where individuality reigns. Popper's (1990) attitude toward macroscopic chance was that it gave rise to interferences that made necessary the switch from forces to propensities at higher scales—the actual fall of an apple depends on far more than the (necessary) force of gravity.

The fact that chance is ubiquitous and unruly does not, however, imply that a system will disintegrate as soon as it encounters a novel event. We recall that feedback controls exist at all levels and work to ameliorate most random events, thereby circumscribing the extent of their effects. In this way, most systems remain immune to the overwhelming majority of complex chance events they encounter. Otherwise, life would not be possible. Of course, a very small minority of aleatoric impacts will be either forceful enough or configured in a particularly damaging way so as to disturb the system; but the integrity of the system, anchored in its controlling feedbacks, is usually able to contain the impact. Here one thinks immediately of the recovery of higher organisms from sickness or injury. Such impacts could induce lasting changes in the operation of the system. Some permanent behaviors may even become necessary so as to anticipate future occurrences of similar disturbances. Immunity in the higher organisms and deciduous behavior in trees are generic examples of such "anticipatory" modes of actions (Rosen 1985b).

A minuscule fraction of disturbances will be capable of driving the system into a new mode of behavior that inflates the

action of one or more self-reinforcing loops at the suprasys-
tem level. Such changes are likely to be rewarded, and the system behavior changes permanently (emerges) as a result. Here
one thinks of speciation, a phenomenon external to conventional evolutionary theory, as an example of such rare, emergent change. Such emergence also lies beyond the domain of
existing stochastic dynamics (e.g., Prigogine 1978) because the
exact configuration of disturbances that can elicit such emergence at any instant jointly depends upon the configuration of
the system itself as well as that of the eliciting disturbance. In
Prigogine's theory of change through fluctuations, all chance
events are assumed to be generically the same. In the complex
world, however, a system will respond to a few perturbations
but remain indifferent to most—just as an individual is vulnerable to particular pathogens and immune to the majority. With
complex systems, it may often occur that the system is responding less to the magnitude of the eliciting disturbance and more
to a lock-and-key correspondence with it, as one observes in
antibody-pathogen interactions. The infrequency of such correspondence would allow types of systems to persist relatively
unchanged over long intervals, only to transform radically its
looks and behavior in response to just the right pattern of disturbance, e.g., the punctuated "equilibria" mentioned at the end
of the last chapter.

That almost all chance is of limited consequence should
prompt the question as to whether the effective domains of all
laws, processes, and rules might be circumscribed as well. By
effective here is meant the ability of a principle to explain relevant
behaviors at the chosen focal level. Of course, we are not suggesting that physical laws are violated at other scales. The laws
of strong nuclear forces, for example, are assumed to continue to
work when one is focusing on macroscopic behavior. (One cannot eliminate, however, the possibility that some laws do not pertain universally. It is debatable, for example, whether the second

law of thermodynamics continues to apply at microscopic scales or whether the first law pertains whenever one progresses from one quantum vacuum to the next.)[2] Engineers have developed a methodology to address multiple scales called *dimensional analysis*, and one of their rules of thumb is that two processes associated with parameters that differ greatly in scale rarely interact (Long 1963). So, unlike physicists, engineers would not spend much time searching for meaningful interactions between, say, quantum phenomena and gravitation—a connection that has eluded physicists for some time now (Ulanowicz 1999a, Hawking 1988).

I am fond of repeating that ecology teaches one humility. If a systems ecologist should formulate a model to describe certain behavior and if that model should happen to contain a singular point (a point where the mathematics becomes pathological), then the ecologist *knows* that, as the independent variables approach the singular point, the model eventually will fail. Any principle proper to ecology will remain valid over only a finite domain of space and time. For ecologists, the universe is packed with "bubbles," each of which delimits the principles and processes endemic to that domain: that is, the universe is "granular" in nature (Allen and Starr 1982).

Dimensional analysis forces the engineer to remain aware of the granular nature of the world. The same does not seem to be true of physicists. Back in chapter 2, in the section titled "A Precipitating Consensus," we remarked how a faith in universality leads many physicists to believe that their models continue to hold arbitrarily close to a singular point. Eventually, the notion of granularity may begin to appeal to many physicists as well. Meanwhile, the finite domains of both chance and necessity argue for dropping universality from the constellation of fundamental scientific assumptions.

Causality by the System

The second postulate explicitly contradicts the Newtonian stricture of closure. In formulating an explanation at a given scale, it no longer is necessary to restrict the search to lower-level phenomena. The most elucidating elements may well lie at the focal level, or even at higher levels (e.g., Salthe 1993, Wimsatt 1995). In chapter 4, we argued that feedback can cause the elements of an evolving system to grow more codependent over time—possibly to the extent that (as is the case with actual organisms) the participating elements lose all capacity to function in isolation. In general, living systems cannot be decomposed and reconstructed at will. It becomes clear that atomism does not prevail over the entire living realm and that it thus should be dropped from the core of basic beliefs.

This last recommendation applies *a fortiori* to notions that attempt to impose atomism in procrustean fashion upon the dynamics of life. We have already remarked on the worn efforts to relate particular traits of a phenome to single genetic alleles (for example, replacing a single nucleotide in the DNA sequence gives rise to a different macroscopic trait in the organism itself, such as hair color). We encounter similar initiatives at higher scales as well. Dawkins (1976), for instance, attempted to generalize the notion of "genes" to apply to larger-scaled systems in the guise of what he called "memes." Memes are the fundamental unit of cultural transmission or the units of imitation that propagate themselves like genes or viruses. Examples of memes are tunes, ideas, catch-phrases, fashions in clothing, ways of making pots or of building arches (Dawkins 1976). Dawkins regards these as the atomistic building blocks of society. But we now realize that these notions can hardly be considered context-free. They all arise in the framework of self-reinforcing cultural feedbacks. They are related to and mutually adapted to other concepts, often so tightly that the thought of their standing autonomously seems

far-fetched. To think of culture as composed entirely of autonomous memes is a truly bankrupt vision.

As we have seen, systems cohere, and in functioning as a
unit their feedbacks often establish a preferred directionality or
telos. Such directionality is purely local and in each case is set
by the particulars of the system. Once a bearing has been established, however, the system no longer reacts blindly to all manner of chance. Random impulses contra gradient to the system's
own internal compass will either be resisted or dampened. One
need only look at social or economic history: The momentum
toward war in 1914 was so strong and involved such a myriad
of social and political structures that the outbreak of hostilities
became inevitable (Solzhenitsyn 1989). The introduction of the
Otto (internal combustion) engine to America rapidly developed its own supporting network of ancillary industries, so
that after a few years, it became impossible to reintroduce the
steam-propelled automobile. Those chance events that reinforce
the system's own tendencies are more apt to induce permanent changes in the system's behavior. Such mutuality between
system and perturbation implies that chance events can possess directions of their own. It is thus inaccurate to speak of
all chance as "blind." Almost all instances of complex chance
exhibit their own local directions. This becomes most apparent
once one interacts with a system possessing its own *telos*. For
example, it could be argued that the contemporaneous building
of more petroleum-cracking facilities, along with an increment
in the manufacture of magnetos and electric starting engines,
prompted and reinforced the ascendancy of the internal combustion engine in the U.S. during the early twentieth century.

Some might object that directionality and feedback (autocatalytic types in particular), in abstraction from atomistic decomposition, render this narrative too mysterious and/or transcendental. The centripetal action by autocatalysis, for example, is totally
foreign to Newtonian systems because nothing can happen in

the linear realm without some eliciting external force. It has long been recognized in dynamics, however, that nonlinear systems sometimes provide an output in the absence of any explicit input (e.g., Guttalu and Flashner 1989). Furthermore, it's not as though mysterious elements haven't long been part of Newtonian (whence gravity and its centripetal action?) and especially Darwinian (the origins of striving and competition?) systems. It's just that the spotlight in conventional narratives remains away from such unexplained features. Hence, I would suggest that our focus upon feedback among inseparable elements is a healthy one that obviates at least as much (if not more) mystery as it introduces.

History Endures

Less mysterious is the third postulate, which has been widely accepted for a long while. According to Darwin, history plays a central role in the evolutionary epic, and it explicitly contradicts the precept of reversibility (his affections for Newtonianism notwithstanding). Even the erasure of history can have important consequences, usually providing freedom to the surviving system to pursue a different course from the one to which it had been constrained.[3] For example, after 746 draws from Polya's urn, the ratio of red to blue balls may have converged to within 3 percent of 0.832. If, for some arbitrary reason, 731 of those balls were to disappear abruptly from the urn, the remaining combination suddenly would have considerably more latitude to converge to another ratio.

Because Noether (1983) demonstrated that reversibility and conservation are two sides of the same coin, it follows that, if reversibility were dropped from the basic postulates, scientists would begin to pay far more attention to nonconservative variables and concepts. Boltzmann began this shift with his definition of statistical entropy, a concept so tightly (and intentionally) associated with irreversibility that it is used far more often with inequalities than with equalities. Boltzmann's direction is continued in ascendency and overhead, both of which build upon his original formula (as reformulated by Shannon [1948]) and which always appear in nonconservative contexts (e.g., increasing ascendency).

It remains an open question to me, however, whether the third postulate is truly independent of the first two. In order to record its history, a system must possess sufficient complexity. In parallel with that requirement, Atlan (1974) demonstrated that a system must exceed a certain threshold in complexity before it can exhibit self-organizing behaviors (as appear in the second postulate). Now, if Atlan's threshold of complexity happens to be higher than the minimum needed to store the system's history, then self-organizing systems would virtually always be accompanied by their histories, and the third postulate would appear redundant. On the other hand, history requires very stable, enduring elements. We have seen how such stability can be the endpoint of cybernetic convergence (Figure 5.5c), but only under very rarefied, simple circumstances. This would suggest that such endpoint configurations are not sufficiently complex to retain their histories, so that the third postulate would be necessary. The choice does not seem obvious, and I leave the question to those more skilled in logic than I am to decide.

While I have made much at every turn of the differences between the Newtonian worldview and the ecological metaphysic, I have not paid as much attention to the discrepancies between process ecology and the Darwinian perspective. In some ways, the new foundation resembles the title of Depew and Weber's book, *Darwinism Evolving*. The ecological scenario, however, differs from conventional evolutionary theory on three major points. First, selection in process ecology is distinctly an internal phenomenon (*sensu* Matsuno 1996), in that a major agency of selection (autocatalysis) acts entirely within the system boundaries. Darwin, to the contrary, was determined to place selection outside that which is undergoing selection (the organism). Second, under process ecology, systems are wont to exhibit a preferred direction proper to their own behavior. This concurs with observations elsewhere of directionality in nature. For example, Schneider and Sagan (2005), Schneider and Kay (1994), Salthe (1993), and Chaisson (2001) all ascribe a preferred thermodynamical direction to the universe. Most neo-Darwinists, to the contrary, remain intent on exorcising any hint of directionality from their discourses (Gould 2002). Third, process ecology holds mutuality to be essential and competition to be derivative of it, in stark contrast to the fundamental position of competition in conventional Darwinian narrative.

Finally, we recall from chapter 2 our remarks on three attributes of the Polya process. We noted that chance, self-reference, and history all played roles in this simplest of artificial processes. That is, the three postulates we have formulated appear to go hand-in-glove with the very idea of process. Popper (1990) seemed to have an intuition of the correspondence when he noted that, to understand causality more fully, we need to think in broader, more general terms: Just as the notion of "force" is but a degenerate (highly simplified) form of something more encompassing (a propensity), a law may be viewed as a precipitated, degenerate remainder of what formerly was a process (viz., Figure 2.3). Darwin's valiant efforts notwithstanding, the notion of process remains foreign to the Newtonian framework. It fits very well, however, into the ecological metaphysic.

New Viewpoint, New Concerns

With process ecology now sufficiently elaborated, we need to assess its potential weaknesses. What, indeed, are its vulnerabilities vis-à-vis the remnants of the Newtonian worldview? Popper (1959) challenged scientists with a virtually impossible task—to attempt with vigor to falsify all hypotheses, including one's own. Recalling how statistical mechanics has been considered for over a century to be the final word that rescued the Newtonian vision (see chapter 2, "Synthesis or Sublimation"), it becomes obvious that we are only too eager to ignore Popper's advice. Still, recognizing that we are never adequate to the task should not deter us from making at least perfunctory efforts at self-criticism.

Perhaps the strongest objection that will be raised against the ecological metaphysic is that it assumes too much of what it will be used to illumine. In particular, too much of the nature of living systems seems to be inscribed in the second postulate—that without the centrifugal assumption of atomism, the deck seems all too stacked in favor of incipient life. Indeed, it is difficult to imagine feedbacks in abstraction from living systems (so long as one includes the designer of artificial situations). But natural, strictly physical entities, such as whirlwinds (Salthe 1993), thunderstorms (Odum 1971), and hurricanes, certainly exist as nonliving manifestations of positive feedbacks. In fact, centripetality (the crux of entification) is so evident in the latter that names are bestowed upon individual hurricanes of sufficient magnitude. So, within the terrestrial domain at least, there does appear to be a purely physical bias favoring what could be labeled proto-ecosystems, and, therefore, progenitors to life as we know it.

But how widespread throughout the universe are the preconditions to life? It is commonly believed that our planet is but one of a vast multitude of others that should possess the necessary requirements, and significant resources are being expended in the search for extraterrestrial intelligence (Morrison, Billingham, and Wolfe 1977). That is as it should be, and the search should continue. But our brief introduction to Elsasser's calculations raises a nagging question about the likelihood of extraterrestrial life. Unfortunately, we do not know how many factors are truly necessary before life can arise (not necessarily life as it appears on earth, but life as a more general dynamical manifestation). If the number of factors is small, say, less than about two dozen, then the probability of life elsewhere is indeed great. If, however, the number is larger, and especially if it is much larger (say, >50), then the uniqueness factor begins to loom, and we should begin to face up to what it might mean to be the one island of life in the vastness of the universe.

Letting Chance out of the Box

In fact, Elsasser's argument was that uniqueness is far more prevalent in the universe than theretofore had been acknowledged. Perhaps the greatest rational problem with the existence of unique chance events is that it makes the radical emergence of characters seem too facile. That is, if singular events do permeate nature, then emergence is one of the most natural things that can happen. (We even had to devote a full chapter to explaining why everything doesn't just disintegrate into uniqueness). I am reminded of a t-shirt that was given to me by a very dear friend, who was familiar with some of my (then nascent) ideas. On the shirt was reproduced a cartoon from *The New Yorker* magazine showing two scientists at a blackboard full of equations. The first is explaining a series of complicated mathematical equations. Midway through the sequence was a box with the statement, ". . . and then a miracle occurs!" The second scientist is pointing to the box, saying, "I think you need to be a little clearer in step 3!"

Not that I am trumpeting miracles. It's just that I am aware of a significant number of colleagues who deny the ontic nature of chance. That is, they doubt that chance events really exist (as breaks in the continuum of causality) and ascribe to them instead an epistemic status—the reflection of our ignorance about details. Given sufficient and precise data, they maintain, laws can still be invoked to give the lie to chance. I have even debated in print whether chance is epistemic or ontic (Patten

1999, Ulanowicz 1999b) and have concluded that the issue is irresolvable by any test that I can imagine. How one decides the issue comes down to a matter of belief. Of course, many scientists grow uncomfortable when someone points out to them that their trust in science rests ultimately upon faith more than self-evident truth, and their tacit desire to avoid addressing belief may account for the relative paucity of works by scientists that critique the fundamental assumptions about nature. The postmodern age, however, has been under way for some time now, and any number of postmodernists on the outside have rushed to highlight the role that belief plays in science. Their criticisms are simply too compelling to be dismissed in cavalier fashion (as was done, for example, by E. O. Wilson [1998] in his book *Consilience*).

It should be evident by now that I regard the existence of chance—especially in its radical manifestations—as absolutely essential to any adequate narrative of evolution. My conviction in this regard is so firm that I would risk admonishing the majority who feel otherwise by paraphrasing Popper's fervor for openness using the same phraseology that Arthur Eddington (1930) did to castigate those unwilling to accept the implications of the second law of thermodynamics:[4]

If someone points out to you that your pet theory of evolution is in disagreement with Fisher's equations, then so much the worse for Fisher's equations. And if your theory contradicts the facts, well, sometimes these experimentalists bungle things. But if your theory cannot accommodate gaps in the causal structure of living systems, I can give you no hope; there is nothing for it but to fall grievously short of providing full understanding of how living systems evolve.

Returning to the problem of reflexivity, do the fundamental assumptions of process ecology subsume some of what they are used to illumine? Yes, to some extent they do. Such reflection is nested in the second postulate. I would hasten to add, however, that this selfsame criticism could also be leveled against the Newtonian metaphysic. So intent were those who helped to precipitate the Newtonian consensus to separate their activities from anything that even remotely resembled the transcendental that they adopted a position so distant from the very natural phenomenon of life as never to be able later to recover it into their ensuing construct. The metaphysic that arose by consensus assumed a completely dead world in what Hans Jonas (1966) has called an "ontology of death." Process ecology, in nowise transcendental, avoids such bias.

A Simpler World Withal

Another perceived shortcoming of process ecology is sure to be that its dynamics are more complicated than those of conventional evolutionary theory. It is indeed difficult to construct a dynamic that is simpler than one finds in neo-Darwinism, a fact that may account both for its enormous popularity and for its "colonial" (Salthe 1989) nature (i.e., the propensity for it to be invoked to explain situations that are of questionable relevance [e.g., Dawkins' memes]). Proponents of neo-Darwinism are all too eager to wield the sword called "Occam's Razor" against any critics, arguing that, all other evidence being equal, the advantage goes to the simpler explanation of events. But, as Albert Einstein was once famously quipped, "Explanations should be as simple as possible and no simpler." And so I raise the concern that perhaps neo-Darwinism is minimalism masquerading as simplicity.

Although that possibility is most intriguing, an even more arresting question is, "Is the neo-Darwinian scenario really as simple as has been touted?" Physicists, for example, caution that we need to consider a problem in its entirety, which includes its particular boundary constraints in addition to its generic working dynamics (Ulanowicz 2004c). Ever faithful to Newtonian tradition, Darwin took pains to place natural selection *external* to the dynamics of selection, which implies that natural selection does its work in the guise of a boundary condition. By moving natural selection to the periphery, Darwin inadvertently diverted attention away from the fact that natural selection in general remains *inexorably complicated*. He focused instead on the more manageable and far simpler dynamics of selection. The ecological scenario, by contrast, places much of the selection as active formal agency *within* the internal dynamics. As a matter of application, the conventional scenario focuses on organisms as the objects of selection. These objects are constantly being impacted by organisms of other types, each of which acts according to some complicated, independent program of its own. Under process ecology, some types become linked in more or less autocatalytic fashion, so that changes in any one of the group conform to a degree with (are rendered meaningful by) the actions of other members of the participating ensemble. Thus grouped, selection at the next level is between a fewer number of "informed patterns of flow" (Wicken and Ulanowicz 1988; see also Krause et al. 2003).

Although such active, internal selection is somewhat more complicated than its neo-Darwinian counterpart, any additional complication is more than compensated by a concomitant simplification of the boundary-value problem. Put another way, what appears as unexplained messiness at the periphery of the neo-Darwinian lens now appears in a meaningful context in the center of the window on process ecology. I am thus led to suggest that the *overall* narrative of development and evolution provided by process ecology is actually *simpler* than the neo-Darwinian scenario.

We have internalized many of what conventionally were considered boundary constraints—in much the same way that evolution has built constraints *into* living systems. In fact, it could even be said that autocatalytic feedback imports the environment into the system, as when the brain of a higher vertebrate images the environment to which the organism has been exposed.[5]

Finally, it should be noted that process ecology rests upon three fundamental postulates—two fewer than comprise the Newtonian foundations. It sometimes happens that one is able to achieve greater generality by relaxing or eliminating particular constraints. For example, by eliminating Euclid's postulate that proscribes parallel lines from ever crossing, the ensuing geometry was able to describe conditions in more complicated spaces, such as on the surface of a sphere.

Reworking the Role of Matter

A third, and possibly the most vigorous, criticism of process ecology is likely to be that it significantly diminishes the active role of material. As we saw in chapter 2, not only did Newtonian closure abrogate formal and final causalities, it also mandated that all efficient causalities issue from material. This was a marked departure from Aristotle's causal typology, where material often appeared in an obligatory but passive role that was acted upon by efficient causes (e.g., bricks and lumber being assembled by laborers, swords being wielded by soldiers, etc). That the very large majority of scientists are deeply wedded to this Enlightenment merger was wryly noted by Richard Lewontin (1997, 31):

We take the side of science in spite of the patent absurdity of some of its constructs, in spite of its failure to fulfill many of its extravagant promises of health and life, in spite of the tolerance of the scientific community for unsubstantiated just-so stories, because we have a prior commitment, a commitment to materialism. It is not that the methods and institutions of science somehow compel us to accept a material explanation of the phenomenal world, but, on the contrary, that we are forced by our a priori adherence to material causes to create an apparatus of investigation and a set of concepts that produce material explanations, no matter how counter-intuitive, no matter how mystifying to the uninitiated. . . .

It is precisely such devotion to material agency that leads many to attribute extraordinary powers to material *objects*. No one laughed, for example, when Dawkins (1976) advanced the metaphor that genes are "selfish" by nature. The media are awash with reports about how particular genes "direct" the development of such-and-such a trait, while absent from such accounts is the network of protein and enzymatic processes that actually read, select, and edit the genome and then implement the subsequent development activity. Such credulity is sanctioned by the ingrained requirement that every legitimate agency be strictly tied to some material object.

Lest the reader recoil in absolute horror, I should hasten to add that process ecology affirms the necessity of the material. As my friend and staunch materialist, Stanley Salthe, defined it for me, "[Materialism] is the (often covert) reliance upon the inherent, primary properties of material substances to foster models that do not explicitly refer to them, but would fail without their implicit presence." That is, material is still required, but its direct action is exerted at scales far removed from the level of relevant agency, which often lies at a higher level. Hence, although process ecology accommodates a "soft" form of materialism, it is a poor fit with the prevailing hard materialism that is preoccupied with forcing an *immediate* connection between agency and material object.

A likely fourth reservation about process ecology is closely related to materialism and was mentioned by Lewontin in the sentence that completed the last quote: "Moreover, that materialism is an absolute, for we cannot allow a Divine Foot in the door."

Of course, one could rightly ask what this statement has to do with process ecology because we have been diligent to avoid any connection with the transcendental. Furthermore, the majority of those with whom I discuss alternatives to neo-Darwinism are avowed agnostics and committed materialists. So where, exactly, is any connection to a "Divine Foot"? Put quite simply, there is no direct connection, but as Popper brought to our attention, we live in a complex world, full of indirect interferences. Physicist

Leonard Susskind (2005) provides a clue to an indirect connection when he wrote what a colleague said in reference to his own theory of "landscapes": "From a political, cultural point of view, it's not that these arguments are religious but that they denude us from our historical strength in opposing religion."

As suggested in chapter 2, in the subsection "Historical Preconditioning," it is hardly surprising that the Newtonian metaphysic served for so long as an excellent tool in the hands of those who opposed religion, for it was perceived (and used) as a weapon against the beliefs that sustained an intemperate clericalism. For many skeptics today, hard materialism lies at the very core of their belief systems, and any departure, no matter how slight, will be perceived as a dire threat to the secular worldview. I should add in fairness that no one should be expected to abandon his or her core beliefs in the face of rational alternatives, any more than I did not abandon mine when confronted by a lucid presentation of secular epiphenomenalism.

Still, I cannot help but question whether an exaggerated materialism is really all that advantageous to the pursuit of science. Although the metaphysic arose out of the palpable need to put as much distance as possible between the activities of the scientist and anything transcendental, it is legitimate to ask whether far more distance was placed between the two than was necessary. The separation became a veritable chasm—an abyss that far exceeded the requirements of methodological naturalism. It could be said of the ensuing gulf that it was so wide that it swallowed any number of perfectly natural phenomena, most notably, life. As a result, we now struggle with the problem of how to approach life starting from an exaggerated Newtonian/ Darwinian minimalism. It is a Sisyphean task, to say the least. So deep is the attachment by so many to a hard materialism that we have rushed to crown as a major success the correlation of genomic forms with phenotypic traits, while still lacking much knowledge of what transpires in between. It is my hope that we can and will do much better, but I believe that a necessary prerequisite to achieving a natural understanding of life is that we abandon the certitude of Newtonian lawfulness and risk accommodating what Prigogine (and Stengers 1984) has called "a world of radical uncertainty."

Dead-End or Gateway?

Nor is the opportunity for deeper understanding of life the only aspect of common experience to be eclipsed by Newtonian ideology. The mechanical worldview has led to the division of society into what C. P. Snow has called "Two Cultures," a bifurcation that Goethe before him had so deeply mourned. An inevitable, rational outcome of the Newtonian assumptions was that all higher human endeavors and sensitivities be relegated to the secondary status of "epiphenomena" (as it perplexed me to learn in my freshman philosophy course). Under the ecological assumptions, however, higher-level functions and systems can now be accorded agency that has been denied them under the Newtonian stricture of closure (Ellis 2005).

I will elaborate on the relationship of process ecology to matters transcendental in the final chapter; suffice it for now to remark that, although the postmodern revolution has not stripped science of all its special attributes, it did open the world's eyes to the role that belief plays in the enterprise. Scientists holding a variety of beliefs (many tacitly) must now defend them on a level playing field against a diversity of assumptions that others from outside their community bring into the larger human discourse.

A New Calculus?

The foregoing concerns about process ecology all relate to philosophical matters, but a significant practical concern remains to be addressed. We note how the way to the Newtonian age was paved by Newton and Leibniz, who independently formulated a highly effective means of treating dynamical systems as continua—the calculus. The material and mechanical attitudes of Hobbes and Descartes had not made much headway until Newton's masterwork *Principia* arrived on the scene. So effective has calculus been as a tool for describing natural phenomena that some have marveled at the amazing correspondence between reality and mathematics (R. Penrose 2005). Calculus indeed has been of amazing utility, but its efficacy wanes as soon as discontinuities appear upon the scene. In order to address discontinuities, some have recently turned to the discrete analog of calculus, now made practical by the advent of modern computers (e.g., Wolfram 2002). This approach, however, is little more than stepping back to the beginnings of calculus (before limits were invoked) and then proceeding under the selfsame Newtonian assumptions.

A somewhat different approach was suggested by Popper (1990), who, in arguing that nature was open, became acutely aware that such a world would require new quantitative methods for its description. His recommendation was to develop "a calculus of conditional probabilities." Popper's suggestion takes on new significance in light of the recent upsurge in the theory of networks (Barabási 2002). As I mentioned earlier, networks of flows of material and energy have a lengthy and significant history in ecosystems theory. Since Lindeman (1942), ecologists have been wont to draw diagrams of "who eats whom" and "by how much?" In the language of network theory, such quantified budgets are called "weighted digraphs." This name itself bespeaks of a nested hierarchy of detail: plain webs of boxes connected by lines are called simply *graphs* (Figure 6.1a). When the lines assume directions (indicated by arrows), the diagram becomes a *digraph* (Figure 6.1b). Finally, when one assigns magnitudes to each arrow, as is done with ecosystem budgets, the diagram takes on the status of a *weighted digraph* (Figure 6.1c).

The majority of networks studied outside of ecology are simple graphs (e.g., Barabási 2002). Within ecology and the social

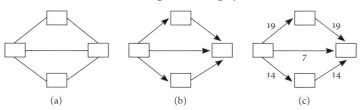

Figure 6.1. The hierarchy of networks: (a) simple graph, (b) digraph, (c) weighted digraph.

sciences, digraphs have been the subject of what is known as "foodweb theory" (Cohen, Briand, and Newman 1990; Cohen et al. 1993; Borgatti and Foster 2003). Food-web theorists have devised a number of indices to characterize the topologies of digraphs representing the feeding relationships in ecological communities. For example, it is sometimes useful to know how well-connected a particular feeding network might be. In a fully connected web with n components, there would be n x (n–1) maximum possible connections. In most ecosystems, only about 15 percent of the feasible connections are realized. One can thus investigate whether the stability of an ecosystem somehow correlates with this connectivity.

One of the first to pay attention to weighted digraphs was economist Wassily Leontief (1951), who employed linear algebra to study cash-flow networks. Leontief took advantage of the fact that any weighted digraph can be represented unambiguously by a matrix of sufficient dimension. One simply places the magnitude of the connection from node i to node j in the ith row and jth column of the matrix. Whence, the matrix corresponding to the 4-component weighted digraph in Figure 6.1c would be,

$$\begin{bmatrix} 0 & 19 & 14 & 17 \\ 0 & 0 & 0 & 19 \\ 0 & 0 & 0 & 14 \\ 0 & 0 & 0 & 0 \end{bmatrix}$$

This one-to-one correspondence between weighted digraphs and matrices allows one to employ matrix algebra to elucidate the properties of the corresponding networks (Ulanowicz 1986, 2004b; Fath and Patten 1999). Chief among the applications has been the calculation of the algebraic powers of an associated matrix to assess how much material passes along network pathways of various lengths. This permits one to conduct an "input/output" survey of the network under study (Hannon 1973). That is, one can estimate how much an input to a particular node affects each of the other compartments, both directly and indirectly. Conversely, one can estimate how much of the activity of a given system component appears in each of the final outputs from the system (Szyrmer and Ulanowicz 1987). In this way, one can trace the effect of, for example, the mining of copper on the production of automobiles. Unfortunately, most economists have not ventured much beyond input/output analyses. (Notable exceptions include Goerner [1999] and Matutinović [2002], who have studied the possible roles of autocatalysis in structuring economic networks.)

Hannon's initial application of matrix algebra to ecosystem networks was subsequently refined and diversified by a number of investigators (Finn 1976, Levine 1980, Bosserman 1981, Szyrmer and Ulanowicz 1987, Ulanowicz and Puccia 1990), notably including the associates of Bernard Patten (Fath and Patten 1999; Gattie, Tollner, and Foutz 2005). Meanwhile, new ways of evaluating trophic levels and other aggregations of complicated webs were developed (Levine 1980; Ulanowicz 1995c; Krause et al. 2003; Allesina, Bondini, and Bondavalli 2005), and algorithms for identifying and quantifying closed cycles within networks were written (Ulanowicz 1983). Many of these various applications have been collated into packages that can be executed on a multitude of computational platforms (Ulanowicz and Kay 1991, Christensen and Pauly 1992, Allesina and Bondavalli 2004, Mason 2006).

Because they are richer in information, weighted digraphs are of greater utility than foodwebs. Bersier (2002), for example, was able to demonstrate that each of the indices defined in foodweb analysis can be reworked so as to apply to weighted digraphs, and the indices created from the latter converge more quickly to and represent more accurately the intended properties. To finally draw the connection between networks and Popper's call for a calculus of conditional probabilities, Alex Zorach (and Ulanowicz 2003) was able to demonstrate a full one-to-one correspondence between weighted digraphs and distributions of conditional probabilities. In effect, Zorach showed that anything one can do with conditional probabilities corresponds to an equivalent operation on weighted digraphs. Hence, Popper's desideratum has essentially been achieved through the development of ecological network analysis.

Matrices are not the only tools useful in analyzing networks. Not long after Lindeman's pioneering work, Robert MacArthur (1955) turned to the burgeoning field of information theory to seek measures of the overall status of trophic networks. He applied Shannon's index to estimate the diversity of *flows* that occur in any particular ecosystem. He had an inkling that this measure might correlate with the degree of stability inherent in the ecosystem that comprised the flows. Unfortunately, MacArthur's emphasis on flows was soon dropped in favor of the conventional preoccupation with objects and stocks, and thereupon followed a full decade of debate over whether the diversity of species populations imparts dynamical stability to ecosystems. The issue floundered for a while and then abruptly and ignominiously collapsed when Robert May (1972) demonstrated the exact converse of what everyone was seeking. That is, using linear dynamical analysis, May showed that a greater number of interacting components was more likely to lead to instability. Ecologists were slow to forget this humiliation at the hands of a physicist (which, of course, only fueled their sense of "physics envy") and thereafter vigorously eschewed all efforts to reintroduce information theory into ecology.

What most ecologists missed by donning their self-imposed blinders against information theory was that, as we have seen, significant advances were then being made in information theory that eventually would allow Rutledge, Basorre, and Mulholland (1976) to reformulate ways of quantifying the overall status of trophic flow networks. As I mentioned in the preface, it was Rutledge's paper that initiated the formulation of ascendency. The behavior of that measure subsequently revealed the lack of a solid Newtonian framework for ecosystem dynamics. Over the following couple of decades, whole systems information indexes akin to ascendency were refined (Ulanowicz 1980; Ulanowicz and Norden 1990; Ulanowicz 2004b, sect. 6) and even extended into multiple dimensions (Pahl-Wostl 1995; Ulanowicz 2004b, sect. 7).

Admittedly, the mathematics already developed to analyze flow networks pales in comparison to the revolutionary contribution made by Newton with his calculus. But the spotlight recently cast upon networks in the frontline journals is certain eventually to spill over into the realm of weighted digraphs. The hope is that someone with mathematical virtuosity will discover methods for analyzing flow networks that will rival the utility of calculus. Such new tools might even permit the parallel description of nature in terms of events and processes. Quantifying unique chance events seems to present a very difficult challenge, but the prospects of constructing a quantitative platform to support the flow dynamics of living systems seem, nevertheless, most promising.

New Viewpoint, New Opportunities

Having discussed some of the potential shortfalls of process ecology, I turn now to possible advantages the new metaphysic may have over previous conventions. I would emphasize first and foremost that the metaphysic consists entirely of natural assumptions. To repeat, there is nothing transcendental in any of the three postulates. In fact, one could turn matters around by arguing what passed through Newton's mind, namely, that universal lawfulness is in itself a transcendental viewpoint. Against that seamless garment, a world full of singular disruptions appears profane by comparison. The framework of the ecological metaphysic fulfills all the requirements of methodological naturalism and should be acceptable to metaphysical naturalists, who actually constitute the majority of those concerned with the inadequacies of the neo-Darwinian worldview. At the same time, the new metaphysic does not bar transcendental extrapolations in the same way that the Newtonian framework did. Some may consider such openness a weakness because, in the words of Susskind quoted above, it denudes them of a weapon against the transcendental. The way I see it, process ecology provides a more neutral theater in which both naturalists and transcendentalists can participate without either having to disdain the intellectual status of the other. The metaphysic has the potential to present science as the intellectual legacy of all of humanity, not just of those who accept a Lucretian materialism.

The Direct Way In

Other than being less encumbered by historical legacy, the three postulates also provide a more direct, logical, and harmonious framework for the dynamics of living systems, which has, after all, been the goal of this treatise. This directness has been achieved by building both chance and circular causality, two nagging enigmas to Newtonian thought, into the primary assumptions about nature. This shift effectively demystifies much of what had remained unspoken or intractable under the conventional narrative, such as the striving that motivates Darwinian struggle, the emergence of novel forms and dynamics, and the origin of life (as we shall discuss presently).

To neutrality and directness one can add that process ecology is more fundamental as well, because the ecological/ evolutionary perspective reveals that we have been reading the text of nature backwards. The palpable material objects of everyday life are actually the *endpoints* of an evolutionary process (Chaisson 2001); they are the equilibrium termini of a more fundamental process, systems whose ascendancies have been allowed to reach their maxima. Historically, we have taken these end stages as the starting points for scientific explanation. It is understandable that science developed this way because systems of unchanging and noninteracting entities are certainly simpler to treat than are mutable, codependent ones. Nevertheless, our progress from this point onward requires that we acknowledge the true order of nature and that we henceforth place the horse before the cart, ontologically speaking.

New Origins?

Putting process ahead of equilibrium obviates another major enigma under the conventional narrative—the origin of life. As Haught (2001) pointed out, a vexing conundrum to the material ontology has been how could life possibly have originated in a world composed entirely of dead matter. One suggestion has been that the precursors of life exist among what only appears to be quiescent matter but actually is already leaning in the direction of life (Chardin 1966). Most approaches, however, focus on dead matter. They begin with simple compounds in retorts that are then zapped with electrical charges (Miller and Urey 1959) or heated (Fox 1995) in the hope that the building blocks of life will appear as a result. Once those units are present, it is assumed they will magically assemble into living entities. Francis Crick (and Orgel [1973]), for example, was so enamored of the molecule he helped to discover (DNA) that he simply could not envision it to be of earthly origin. He concluded that it must have arrived from extraterrestrial sources. Once on earth, the polymer somehow impressed a whole host of other polymers and enzymes into its service. These approaches, with their Eleatic emphases on objects as the origins of agency, are akin to expecting Ezekiel's dry bones suddenly to put on flesh and begin to dance. The ecological metaphysic, however, follows a Milesian approach. In particular, process ecology suggests that the agency that creates living entities is to be sought among configurations of ongoing processes.

Howard Odum (1971) followed this path when he proposed that proto-*ecological* systems must already have been in existence before proto-*organisms* could have arisen. In his scenario at least two opposing (agonistic) reactions (like oxidation-reduction [Fiscus 2001]) had to transpire in separate spatial regions. One volume or area had to contain a source of energy and another had to serve as a sink to convey the entropy created by use of the source out of the system. Physical circulation between the two domains was necessary. Such a "proto-ecosystem" or circular configuration of processes provides the initial animation notably lacking in earlier scenarios. We have seen that circular configurations of processes are capable of engendering selection and that they can naturally give rise to more complicated but smaller cyclical configurations (proto-organisms). Such transition poses no enigma: in irreversible thermodynamics, processes are assumed all the time to engender (and couple with) other processes. (It should be noted that one form of change begetting another is fully consistent, dimension-wise.) Large cyclical motions spawn smaller ones as the normal matter of course— as, for example, when large-scale turbulent eddies shed smaller ones. Likewise, the cosmological cascade begins with large-scale galaxies that spawn stars within themselves, which, in their turn, create the heavy elements necessary for life, etc. Corliss (1992) has suggested that a scenario like the one described by Odum might have played out around archean thermal springs, an idea that recently has found new enthusiasts in Harold Morowitz and Robert Hazen (Cody et al. 2001). So, process ecology, wherein the origins of new objects are mediated by configurations of processes, provides a more consistent framework for exploring the origin of life than heretofore has been available.

Still Earlier Origins

As I have hinted, contemporary physics describes the evolution of matter in terms that markedly parallel the development scenario portrayed by process ecology (Chaisson 2001). After the initial big bang, a very subtle asymmetry (one in 10^9) led to the formation of more matter than antimatter. Contemporaneous with the formation of equilibrium material forms, nonlinearities (feedbacks) arose in space/time to separate the strong, weak, and electromagnetic forces from the more generalized force (referred to as GUT). Through successive feedback, forms grew quite stable on their own, and their interactions (forces) grew precise, until the physical world with its accompanying laws and constants eventually took shape.

The scenario of coalescence can be thought of as akin to the system ascendency approaching its limiting equilibrium configuration, as depicted in Figure 5.5. The reader might be puzzled as to how ascendency, which requires differentiated species, could be applicable to phenomena wherein types have not yet arisen from the common medium. It happens that differentiation can be accomplished via spatial separation. In this way, the ascendency of a homogeneous medium moving over a spatial domain can be reckoned in a straightforward manner (Ulanowicz and Zickel 2005). First, the spatial domain is divided into finite segments, usually a gridwork of spatial elements. Because the division of the finite spatial domain is into a finite number of compartments, those compartments can all be numbered serially in any convenient manner. As medium passes from one grid box, i, to the next, j, it can be tallied as a flow from i to j, or T_{ij}, to use the notation defined in the last chapter. The magnitudes of the ascendency and overhead of the flow field can then be calculated using all such T_{ij}. The resulting ascendency represents the organization inherent in the power of the flow field. Although the absolute values of ascendency and overhead thus calculated will depend on the scale of the imposed grid, relative changes of the variables over time will portray whether the flow field is becoming more or less organized, etc.

In this scheme for representing flow fields, the counterpart to the equilibrium, equiponderant cycle in Figure 5.5c is a perfectly circular vortex (Ulanowicz and Zickel 2005). One can thus depict nascent material as a mix of such vortices, which are constantly being destroyed by interaction with the surrounding sea of photons or created by collisions between photons themselves. As space in general expands after the big bang, the photons diminish in energy (their wavelengths dilate), so that collisions between photons and vortices become both less violent and less frequent. Eventually, the vortices effectively become isolated from the photon gas and take on a prolonged, self-contained (or equilibrium) existence (as in Figure 5.5c).

The inference to be drawn from this scenario is that even what appear to be quiescent, unchanging material objects derive from and remain, in essence, configurations of processes. As Mark Bickhardt and Donald Campbell (1999, 331; Hoffmeyer 2008) wrote,

The critical point is that quantum field processes have no existence independent of configuration of process: quantum fields are process and can only exist in various patterns. Those patterns will be of many different physical and temporal scales, but they are all equally patterns of quantum field process. Therefore, there is no bottoming out level in quantum field theory—it is patterns of process all the way down, and all the way up.

The take-home message is that both matter and life have likely proceeded from a common dynamic, and one can follow these developments in terms of ascendency and related variables. More importantly, the process of increasing ascendency as induced by augmented feedback is seen to be more basic and fundamental than either material or life, both of which are seen to share a common dynamical legacy. Thus depicted, the arrival of life on the scene was no more (or less) exceptional than was the earlier appearance of matter, and the necessity to explain how dead matter suddenly becomes animated simply vanishes (Ulanowicz 2002).

7

The View out the Window

Βλέπομεν γὰρ ἀρτι δί ἐσόπτρου ἐν αἰνίγματι
[We still see obscurely, as in a mirror.]
—Shaul of Tarsus

Newtonian Chains Unbound

Having labored to frame a new window upon nature, it is appropriate now to step back and ask what has been gained from our labors? What new insights can be glimpsed through this third window? Well, for one, we have just encountered a fresh perspective on the origin of life—the proto-ecosystem. But there is still more for the looking.

Science versus Culture?

One salubrious potential for process ecology is to contribute toward healing the rift between Snow's "two cultures"—science and the humanities. Peirce, Popper, and Prigogine all argued at length against the belief that the universe is closed because closure effectively trivializes any creativity that arises out of the arts and the humanities. If the human being were an onion and all human activity merely epiphenomena, then the arts would necessarily remain subsidiary to science. Visions of achieving a "grand unified theory" in physics and the ostensible triumph of liberal welfare economics across the globe have prompted some to talk triumphantly about a possible "end to history" (Fukuyama 2002), as if history could be taken seriously in a clockwork universe. Newtonian science has been used (at times not without justification) for deflating the exalted position that humanity had built for itself. But, with the process vision, we now come full circle and invite science to step down off its pedestal to mix as equal with other human endeavors.

Openness levels and frees culture at large, while liberating the individual as well. In a clockwork world, humans are but automatons, appearances to the contrary notwithstanding. Human thought is but the movement of electrons in the brain (Churchland and Churchland 1998) that plays out according to the laws of nature, which eventually will be used to map those mechanics in detail. In other words, the brain is but a machine (Franklin 1983), and no such thing as "free will" exists. Some have struggled against such suffocating confinement, arguing that the existence of ontic chance among the netherworld of atomic phenomena provides openings among the brain's electrons through which volition may enter (e.g., R. Penrose 1994). That is, indeterminacy among the quantum dimensions in the brain may ramify to higher levels, thereby freeing thought from the determinacy of law. But we have seen that there is no reason to confine chance to the microworld. In the open world of Popper and Elsasser, the aleatoric arises at *all* dimensions, including the macrodimensions of brain circuitry and biochemistry.

Furthermore, universality remains foreign to the world of process ecology, which appears granular in all directions and dimensions. This new perception of the looseness among the layers of phenomena applies as well to the human brain. Several levels separate the firings of neural synapses from the higher-level, slower cognitive functions that go into making decisions (Juarrero 1999, Murphy and Brown 2007). Moreover, those larger-scale functions are innately coupled with the external world and, in the case of humanity (and possibly with some

higher mammals) with culture at large. Yes, electrons necessarily move with each thought one has, but no longer may the
influence of larger events on the patterns of those movements
be ignored. The metaphor of the muscadine grapevine suggests
that higher-level functions may exert suprafacience by reworking the structures of lower-level neural networks (Murphy and
Brown 2007). The belief that bottom-level neural firings fully
determine higher-level outcomes becomes fatuous.

Pointing the Finger

The revival of free will imparts renewed credibility to the notion
of personal responsibility. Everyone agrees that organic dys-
function can give rise to any number of pathological and crimi-
nal behaviors, but, in an open world, it never will be possible to
trace all individual behavior back to defective genes or aberrant
neuronal structures. Those who are discomfited by this revival
of personal culpability may find some solace, however, in the
realization that the new metaphysic allows one to apportion
responsibility to higher levels as well.

In earlier work (Ulanowicz 1997), I pointed to the calamity
that befell the Challenger space shuttle, which blew up several
minutes into flight on January 28, 1986. In the aftermath of the
misfortune, a commission was formed to evaluate NASA, its
operations, and personnel down to the minutest detail. It was a
good search conducted in fine Newtonian fashion, and, eventually the culprit O-ring was identified. I remarked in chapters 1
and 5 how important it is to recognize the agonistic natures of
efficiency and reliability. I don't recall an account of this conflict
affecting the Challenger investigation in any prominent way. Had
this complementarity been considered, some blame might have
been attributed to the larger social and political system. It is noteworthy that, prior to this catastrophe, no other American had
died in space flight (although three astronauts did perish in a fire
during ground tests), a record that owed to intensive efforts since
early in the NASA program to ensure flight safety through mas-

sive system redundancy. As regards efficiently sending a person to a given point in space and back, however, such redundancy will always appear as overhead. Economic times were not as flush preceding the disaster, so Congress and taxpayers were vigilant to trim expenditures anywhere they could. The push was on to make each government program as efficient as possible, and this message was transmitted to NASA in any number of ways. And so, reliability was tacitly deemphasized, even as the machines themselves were growing more complex. But we now know that, in order to persist, a system must retain a modicum of *both* efficiency (ascendency) and reliability (overhead). By bleeding overhead, the tragic outcome became simply a matter of when and how. Significant culpability in the Challenger case legitimately belongs to society as a whole.

Homo cum Machina?

We have argued that genes and neural firings cannot be made fully to blame for any action because the brain and the body are categorically not machines. Yet the reigning metaphor in Western medicine has for some time now been *corpus cum machina*. To be sure, the metaphor has served well at times, especially in fields like cardiovascular therapy, where the analogies of heart and vessels to pumps and pipes are not entirely off the mark. It's a little more difficult, however, to envision cancer or diabetes mechanistically, but a significant fraction, if not a majority, of support for cancer research goes toward investigating mechanistic etiologies, such as the identification of specific "oncogenes." The worry is that an exaggerated confidence that human physiology is genetically driven could divert needed attention from the focus that cancer is fundamentally a system-level disease (Sonnenschein and Soto 2006). Three out of four events necessary before cancer can metastasize transpire at mesoscopic levels of the body system (escape from local tissue, evasion of surveillance elsewhere, and invasion of new tissue type).[1]

To be sure, these system-level checks on cancer all involve

genomes in one way or another, and knowledge of those roles is surely necessary in order to make progress against the disease. What I find worrisome, however, is that many are attributing active agency to the genomes because interpreting them as efficient causes gives rise to a "silver-bullet" mentality that focuses attention upon what are largely passive elements. If one demotes genomes to the status of (passive) material causes, one does not thereby abrogate the necessity to include them in any analysis of the pathology. The insufficiency of material cause as full explanation, however, motivates the investigator to continue the search for efficient agency among the enzymatic/proteomic system of reactions that interface with and interpret the genomes. By thereby widening our search, we are more likely to encounter effective treatments, which might include system-level therapies such as those that involve the immune system. Fortunately, the idea is dawning in some quarters that much of the agency at work in physiology in general resides in networks of metabolic reactions (Strohman 2001), which also appear to be the key agencies at work in the deployment of genomes during ontogeny (Coffman 2006).

Of all biotic ensembles, perhaps the immune system of higher animals most resembles the ecosystem (Vaz and Carvalho 1994). Studying the immune system with the methods of ecological network analysis no longer seems far-fetched. Again, many current disease therapies focus upon eliminating the "silver bullet" that initiated them. But more effort could be spent pursuing therapies that involve the entire immune system.

Interest has grown over the last two decades in the hypothesis that a patient's psychological state can significantly affect his or her recovery from pathology. This highly un-Newtonian pursuit seems to be gaining credibility on purely phenomenological grounds. It is now well established that mental state can alter the immune system (Rabin 1999), a fact that possibly serves as an explicit example of suprafacience by higher-level functions.

Beyond the Pale

Ameliorating Metaphysical Impatience?

Morality and psychology bring us perceptibly closer to spirituality. While I have studiously endeavored to follow all the strictures of methodological naturalism, I have also made no secret of the fact that I sympathize with the secularist Bateson, who complained that our conventional wisdom bars all access to the sacred. Bateson's notion of the sacred was avowedly non-transcendental, but, for myself and possibly for some readers, the cultural grades imperceptibly into the transcendental. I also am aware that most readers probably would rather not venture further in the direction of the transcendental, and I respect that reticence. Nevertheless, the dialogues between theists and naturalists and between theologians and scientists are very much in the headlines these days, and I would feel remiss did I not mention the relevance of process ecology to those controversies because I believe the new formulation offers significant hope for ameliorating or obviating some of the perceived conflicts.

That the conversation between scientists and theologians is so topical owes in some measure to recent monetary support by the John Templeton Foundation, but, in substance, it has been prompted by the postmodern critique of the privileged role that science plays vis-à-vis other ways of knowing. Because discussions at this interface often relate to the very core of a participant's self-image, they can become animated and, unfortunately at times, vehement. Such emotion derives from the distrust individuals often hold for those expressing opposing opinions. Not infrequently, there exists the fear, too often justified, that one's opponent is seeking to "seize one's philosophical position" (Haught 2000). Thus, theists, at least since the time of Thomas Huxley, have felt that some in science wish to use that body of facts to extinguish any possibility of belief in agency beyond the natural. Susskind's comment in the last chapter revealed that such worry persists and is not outright paranoia (e.g., Dennett 2006, Dawkins 2006). More recently, metaphysical naturalists have felt themselves besieged by many in the public who seek to make part of public education the belief that an Intelligent Being beyond human capacity is required to explain the order in nature (Behe 1996, Dembski 1998). Either of these conflicting intentions is an example of what theologian John Haught (2000) calls "metaphysical impatience"—a desire to establish the certainty of one's beliefs and to extirpate opposing convictions, an unwillingness on either side to admit that we inhabit, as Ilya Prigogine (and Stengers 1984) put it, a world of radical uncertainty.

Radical uncertainty among individuals is the result of ontological indeterminacy in nature, which is written into the first axiom of process ecology. It should come as no surprise, therefore, that the framework proposed in this essay can serve to dampen the temptation to engage in metaphysical impatience. At least such is my hope. Like some other theists in science, I have always remained keenly aware of an internal, personal dialogue between belief and doubt (naturalism). One of the benefits of rehearsing this dialogue in one's mind is that doing so soon makes it impossible to deride one's opponent, regardless of which side one takes in any particular discussion on the topic. And this internal dialogue has in large measure been why I have endeavored to produce a framework for viewing the world that by its very nature discourages metaphysical impatience and encourages respect for all responsible expression of belief.

At the risk of trying the reader's patience, I repeat how I have sought always to pursue scientific research under the tenets of methodological naturalism. That is, I have put all my ideas to the strict test that they involve only the natural (even if the delimitation of that term should at times become problematical [Brooke, forthcoming]). There are no skyhooks in what I propose, Mr. Dennett should be pleased to learn. At the same time, I have felt it necessary to push the boundaries of science beyond what some would deem advisable. For example, I mentioned in the preface that I eschew epiphenomenalism and encourage scientific engagement with macroscale agency wherever possible. As I hope this essay has communicated, I regard striving, volition, and emotion to be agencies in the play we call nature and legitimate subjects for scrutiny by science (albeit not of the Newtonian ilk). I am not above asking "Why?"; in fact, I question whether much progress would ever have been made had this query remained off limits to scientists. Besides, answers are sometimes forthcoming, as I hope my exegesis of striving and competition might have suggested.

Some theists among the readers may see the imposition of methodological naturalism on oneself as unnecessary. They may even worry that a completely naturalistic metaphysic only repeats history by methodically excluding mention of the Divine in nature. But I believe that such fears arise largely out of experience under a legacy that intentionally had kept nature closed. The ecological metaphysic, by contrast, is anything but closed, and I would suggest that the newly discovered rational necessity for openness provides hope for all who would wish to balance a belief in the transcendent with a sober view of the natural world. I would think that the possibility of such equanimity would be welcome news to those young researchers whose beliefs may have been shaken by the conventional image of a world that is wholly determined by laws, a universe in which there is nothing left for a creator (or a creative scientist, for that matter) to do. Or, as Philip Hefner (2000), founder of the Zygon Institute for Religion and Science, once lamented, "There is no 'wiggle room' for God to act in the world." I can only hope that I have presented an alternative vision that is wholly natural and intellectually coherent, one that provides a parallel and ultimately more fundamental approach to the wonders of nature without impeding access to the sacred. I hope as well that the alternative ecological metaphysic will encourage young theists who enter science that they may remain believers without compromising their rationality. Contrary to the opinion of some of their prominent older colleagues, when they enter a house of worship, it is not necessary for them to "check their brain at the door." At the same time, because the ecological narrative remains wholly natural, theists are obliged to respect the beliefs of metaphysical naturalists who eschew any connections with the transcendental.

The Almighty or the Aleatoric?

It is certain that naturalists and theists will continue to disagree about the exact limits to agency. Not only will the controversy be unavoidable, it will (in my opinion) also remain irresolvable. The more encouraging news is that, under process ecology, several of the other controversies between naturalists and theists will either vanish or moderate substantially. The issues of free will and the origin of life are two examples of how process thinking can mitigate old disputes. The same potential for resolution and/or moderation applies, I believe, to issues having even greater theological content, like, for example, the question of immanent divine action. As already noted, the prominence of the postulate of closure led many to believe that each and every event and outcome in the observable world was determined by Newtonian-like law. If a god were to exist in such a world, it would be fundamentally inconsistent for the deity to violate its own foundational laws. That is, the only role a creator could possibly play in a totally lawful universe would be to supply the initial conditions and provide the impulse for the ensuing script. Belief in such a god has been called deism, which became popular in the wake of the Enlightenment, and even attracted several notables among the Founding Fathers of the American democracy (Wills, 1978).[2]

That God would intervene in the subsequent unfolding of

the natural order was inconsistent with the deist viewpoint. Miracles have no place in the deist scenario: there simply isn't enough "wiggle room" for a deity to act. Corollary to this belief is the conclusion that intercessory prayer is futile. Against such a background, Dostoyevsky marveled at the faith of the widow lighting a candle and praying for God to heal her sick son.

Without belaboring the point, it should be abundantly clear by now that process ecology is rife with "wiggle room." Should a god wish to intervene in response to a human petition, process ecology would provide an abundance of holes in the causal fabric via which to do so—if necessary in coordinated fashion and across a hierarchy of levels. Some will rush to dismiss this assertion as another "God-of-the-Gaps" argument, whereby gaps in scientific *knowledge* are prematurely ascribed to the action of God (an example of metaphysical impatience). But John Polkinghorne has cautioned us that there are epistemic gaps of ignorance in contrast to ontological gaps of necessity (Davis 1998), by which he meant that the gaps in our understanding about how things happen eventually will be filled by new, emerging scientific facts and theories. In such cases, we are justifiably cautioned against impatience. But there also exist gaps that are part of the formal structure of science, which reflect the ontic openness of nature. Examples include Gödel's incompleteness theorem, Heisenberg's uncertainty principle, the Pauli exclusion principle, and Elsasser's unique events. To deny these latter gaps is to defy rationality. Furthermore, such necessary gaps make it impossible to preclude immanent divine action.

Intelligence or Impatience?

Like Occam's razor, process ecology is a double-edged blade (Ulanowicz 1995a). We see how the first postulate provides abundant wiggle room, but, by the same token, the second assumption allows for events to cohere and *fills in gaps* (Ulanowicz, forthcoming). Regarding such gaps, the theory of Intelligent Design

(ID) has made big news in the United States as an alternative that many wish to have taught in schools alongside Darwinian evolution. As part of ID, William Dembski (1998) identifies "complex specified information" (CSI), which he claims is conserved and cannot be created via the mechanisms that comprise neo-Darwinism. CSI, therefore, must be the work of a Creator. The first problem with accepting Intelligent Design is that one must first buy into the mechanistic framework that supports neo-Darwinism. Insofar as Behe and Dembski remain within the orbit of Newtonian dynamics, they simply are reflecting Bateson's concerns that it is difficult, if not impossible, to create information. There are, however, no obstacles to creating information within the more open realm of process ecology, as readily happens whenever the ascendency of an ecosystem increases. That is, under the ecological metaphysic, which sets radical chance against autocatalysis, nature appears fully capable of doing what Dembski et al. said is impossible under clockwork assumptions. Research built around process thinking shows promise for filling in the epistemic gaps identified by ID.

Nevertheless, much of the ridicule that has been leveled at Dembski and associates has, in my opinion, been gratuitous. There is much to learn by thoughtful deconstruction even of mistaken arguments. Concerning their methodology, Dembski employed exactly the same informational calculus that Rutledge had introduced to ecology some thirty years ago and which I subsequently used to parse ordered complexity from its disordered counterpart (ascendency from overhead). It's just that Dembski erred (in my opinion) by identifying the ordered component in the decomposition as the irreducible complexity of interest. Elsasser's irreducible (and disordered) complexity resides in the complementary term that Dembski chose to ignore.

The Necessity of Spilt Milk

The first postulate incorporates the ubiquitous irreducible complexity that serves to disrupt the workings of systems. In social and moral terms, this is the problem of suffering and evil, and how to reconcile the notion of a providential God with the presence of evil and suffering has always posed a major problem for theologians and for believers in general. The problem has even been given a name, *theodicy*. Under the framework of process ecology, the conundrum of theodicy certainly does not disappear; rather it takes on a different, slightly less mysterious form. As we remarked earlier, the transactional nature of the world entails that disturbance is essential if systems are to evolve. Among such disorder are many that cause suffering (chance) and some that could be labeled as evil (interference). As regards chance, one reads daily about examples of suffering that have engendered beneficent action, such as the death of a close relative that motivated a student to become a successful medical researcher.

Intentional evil is harder to accommodate, although it obviously cannot be eliminated without revoking free will (and freedom is a prerequisite for love). A second consideration arises out of scholastic metaphysics, which was clear in allowing that good could result accidentally from evil (Aquinas 1981). One of my research collaborators was born in Hiroshima not too many years after that city had been destroyed by a nuclear weapon at the end of World War II. I once was moved to ask him how he felt about those who had ordered that his city be bombed. He replied that his father had been badly burned by the blast, and was attended by a young nurse, whom he later married. Without the bomb my colleague would not exist. Of course, enormous asymmetry in the magnitudes of catastrophic cause and beneficial effect render this example problematical, but situations of the opposite imbalance lie perhaps closer to the point.

Albert Einstein, for example, deviated from his assigned duties as clerk in the Swiss Patent Office to spend most hours of his days working on what was to become the theory of relativity, one of the cornerstones of our first window on the universe. (To complicate matters even further, we note as how relativity theory was instrumental in our gaining access to nuclear energy and weaponry.) It soon becomes obvious that we inhabit a world in a moral tangle, but our insight into the transactional nature of development makes it clear that, without tolerance for some minor evils, humanity could not progress and would risk collapse. Without mistakes, there can be no creativity. Suffering and petty evil become a *necessary* part of the evolutionary scenario (Ulanowicz 2004c; Keller 2005; Jackelén, forthcoming; Callahan 2003). Theodicy obviously remains a dilemma for the believer, but process thinking now allows us to regard it from a different angle. The problem no longer concerns ontology; but, as the existence of my colleague from Hiroshima suggests, it becomes one of magnitude—why would a provident God allow *major* suffering and evil to persist?

An Epistemological Veil

The juxtaposition of divine action alongside that of theodicy prompts the believer to ask whether the finger of God lies behind every chance event. Evidently, some theists believe that it does, in which case theodicy for these individuals resolves into finding God's larger design behind every major evil. Another perspective is possible, however. Some believers are willing to accept that no higher agency lies behind most instances of chance. These things happen, as the saying goes, and constitute a necessary part of the natural dynamic. As believers, however, they depart from naturalism in maintaining that the action of God may lie behind *some* apparently stochastic events, especially those connected with the human/divine dialogue (as recorded in scripture) or those which promote some obvious good (e.g., the concept of grace). Philip Hefner (2000, 61) worried that such "selective divine determinism" might be unacceptable, because it appears to render "both nature and God unreliable." But nature, as seen through the lens of process ecology, is not wholly deterministic. It nevertheless manages to incubate reliable elements. Why, then, require automatic behavior of a deity who does not impose determinism upon the universe?

In the end, neither theist nor naturalist has any scientific way of resolving whether or not higher agency lies behind a given singular event. The question equates to whether chance is ontological or merely a disguise (Ulanowicz 1999b). The first postulate accepts chance as ontological. Regardless of how one decides this issue, all who venture to look out the third window perceive in the distant background an "epistemological veil" that neither rationality nor empiricism can penetrate. Whether that veil shrouds emptiness or obscures a higher agency remains perforce unknowable. Radical uncertainty is the portion of *all* who inhabit this mortal coil. In such a world, patience with one another becomes a clear virtue.

Counting Conceptual Shifts

Indeterminacy, the crux of the second law of thermodynamics, gradually erodes history. So before our vision dims by staying too long afield of methodological naturalism, it becomes advisable to pause yet a second time and take stock of what frames the third window on nature. A proper understanding of process ecology requires several shifts in emphasis regarding how the world operates. The primary adjustment necessary is to pay less court to deterministic law and pay more attention to process. Process is not something new to science, but mostly it has remained obscured in the penumbra of matter. History, likewise, has been an element of science since before Darwin, but our desire to predict and control nature has kept it in the shadows as well.

Process is correctly seen as the generalization of law. That is, in the evolutionary course of the universe, laws emerged out of inchoate processes eventually to become static, degenerate forms of the latter. By analogy, Popper has suggested that we consider propensities to be generalizations of physical forces. Popper's shift helps us to make better sense of complex situations because identifying interdependent propensities as active agencies can explain more than can our traditional reliance on isolated forces. That is, we are better off in a complex world diverting some attention from objects toward configurations of processes as the key agencies that impart pattern to the world. Physicists surmise that, very early in the big bang, the interaction of processes (photons) gave rise to the first objects (paired matter and antimatter), and the ontological priority of process over object continues into to our present world.[3] It was an accident of history that agency was first identified with objects. One goal of this presentation has been to restore the priority of process over objects in how we see the natural world. That configurations of processes may give rise to and remain embedded in behaviors that had heretofore been disregarded as epiphenomena (e.g., striving by organisms) is an exciting new feature of looking out of the third window.

One advantage of focusing on configurations of processes is that they interact with chance in creative ways not possible through the conjunction of simple forces. As a result, chance no longer appears a threat to the scientific enterprise, as it once did. We are free now to accept it as an essential part of nature and to acknowledge it as a source of true novelty out of which new forms emerge quite naturally.

This schema wherein patterns arise out of the interplay between chance and process surprisingly rests upon fewer crucial assumptions than did the Newtonian project. Thus, process ecology is to the Newtonian construct as non-Euclidean geometry is to the classical version—the former both rest upon fewer axioms, and they expand the horizon of applications. Furthermore, each of the postulates of process ecology is the antithesis of a counterpart in the conventional metaphysic. It is as if the world as seen through the first two windows had heretofore been inverted but now appears aright in the third.

Crossing the Divide

If, indeed, the Newtonian image has been an inverted one, how is it that the conventional metaphysic not only endured for so long, but over its time has led to so many stunning advances and insights? I suggest in the first place that the human mind has significant capacity to adjust to distortion. In psychology, for example, when a subject is fitted with eyeglasses that invert the image reaching the retina, he or she quickly learns to compensate and soon grows unaware of the upended perspective. More to the point, however, I had proposed earlier that the Newtonian vision was very convenient for solving many of the simpler problems that nature presents. But the matter goes beyond mere simplicity. Rather, I think the Newtonian image remains appropriate in its own domain, but fails outside that realm. (Nature is granular.) Relevant here is the organic fold described in chapter 5, the subsection titled "The Organic Fold." The reader will recall that the existence of the fold was uncovered by the observation that the directions in which causal action are exerted tend to be reversed at opposite ends of the organizational hierarchy. Causes act primarily bottom-up at microscales, whereas top-down influence provides more relevant explanation at higher levels. Qualitatively, events and patterns on opposite sides of the fold differ in two respects—(a) dynamics and (b) history. Dynamics below the fold are characterized by objects moving at the behest of simple forces. Above the fold, they resemble more a transaction between feedback and complex chance. As for history, it goes largely unrecorded among the simple entities below the fold. The past can be adequately recorded only in such complex configurations as exist above the fold.

While matters at the extremes of the hierarchy remain relatively clear-cut, the transition across the fold becomes ambiguous terrain. Just on the physical side of the fold there exist active configurations, such as whirlwinds (Salthe 1993), tornadoes, thunderstorms, and hurricanes (Odum 1971) that develop in a fashion similar to living systems, save that they remain incapable of recording their own histories. Immediately above the fold are structures like viruses that are capable of storing their own phylogenetic legacies but which must capture the dynamics of other living forms in order to propagate. Only those entities that can incorporate both the dynamics *and* the stored history as prescribed by all three ecological postulates can be considered truly alive.

Such ambiguities notwithstanding, the existence of the fold implies that living entities are qualitatively discontinuous from nonliving ones. Most will consider this a seriously heterodox view that hearkens back to ancient attitudes and contradicts the received modern wisdom that no qualitative difference exists between the quick and the dead. Here it helps to note that, if one is preoccupied with objects and the differences between them, then the transition between the living and the nonliving remains somewhat obscured. By focusing on things, one's attention is diverted from what is most indicative of life. It is only by adopting a process standpoint that the demarcation emerges into view. With one's eye on processes, a live deer looks qualitatively very distinct from a dead deer.

In retrospect, we can now ask why we were ever led to believe that no qualitative divide separates the living from the nonliving. Is not discontinuous qualitative change very much evident among physical systems? One sees numerous phase transitions between different forms of physical matter. Why then forbid any discontinuity between the even more dissimilar realms of life and nonlife? The answer seems to involve ideology more than rationality: to allow such discontinuity would be a de facto admission that the dynamic of Newtonian forces acting on dead matter is wholly inadequate to the task of building life.

To me, it was precisely the popular denial of the organic fold that most troubled Gregory Bateson. A rationalist agnostic, Bateson nevertheless couched his objections to the reigning scientific wisdom by complaining that it "denied access to the sacred." Of course, he regarded the word *sacred* in abstraction of any transcendental or religious connotations. But what he was suggesting was that the denial of any qualitative shift in dynamics as one passes into the realm of the living implies that life is in essence every bit as simple as nonlife, and Bateson was fervently convinced that most actions spawned by this belief will prove ultimately self-destructive. As antidote, he adopted a dualist perspective that effectively used the organic fold to separate his "pleroma" from what he called "creatura." His deliberate use of the latter term was meant to signify that the dynamics of life escape the Newtonian orbit and must be considered on their own. Finally, in order to achieve his fuller vision, Bateson pointed us in the direction of ecology, an intuition that I believe was truly prophetic.

I believe that metaphysical naturalists and theists alike would benefit enormously by heeding Bateson and opening one's mind to the possibility that life deserves more than an ontology of death. For scientists to admit such a prospect is to open a window upon a whole new approach to interpreting and acting upon the world around us. It would be like removing the rose-colored spectacles of mechanism in order to behold anew a world of dazzling color. Confronting agency more fundamental than matter itself should exhilarate and inspire us to rework and renew our entire social, artistic, political, and economic ethos in the light that it is living creatures and not molecules that transform the immediate world. Most importantly, by adopting process ecology, we are abandoning the prevailing "cosmology of despair" (Haught 2000) and embracing instead a vision of hope.

Notes

Preface

1. The alternative spelling *ascendency* is used throughout this book to distinguish the concept from the usual meanings of the term *ascendancy*.

Chapter 1: Introduction

1. It is worth noting here that quantified, directed networks had been the object of much interest in ecology for more than a half-century before far simpler networks recently were "discovered" late in the 1990s by physicists (e.g., Barabási 2002).

2. Lest mention of the "R word" make some readers uneasy about where this thesis is going, I would like to assure skeptics that the development of the subject up until the final chapters will remain wholly within the confines of methodological naturalism, eschewing any mention of the transcendental or the supernatural.

Chapter 2: Two Open Windows on Nature

1. The Balmer series describes the ratios of the frequencies of spectral emissions from the hydrogen atom. A binary number is the most elemental way to represent a number (base 2) and can be written using any arbitrary representation of the integers "one" and "zero."

2. One can intuit such possibility simply by observing the disparity between Western and Eastern approaches to medicine made by essentially identical beings. While Westerners have previously dismissed Eastern medicine, a new appreciation for Oriental therapies is now dawning.

3. The headlines of today's newspapers reveal that this condition still prevails in some regions outside the Western world.

4. The author is indebted to Robert Artigiani for bringing these events to his attention.

5. Some prefer that the term *atomism* should refer solely to the process of identifying the parts and that disaggregation and recomposition bear instead the title *decomposability* (Andrzej Konopka, personal communication).

6. This is a consequence of the fact that replacing time, t, with its negative, –t, leaves the equations that describe the four laws of force unchanged.

7. Had Darwin aspired instead to become the "Leibniz of a blade of grass," matters might have evolved quite differently, possibly in the direction which we are headed in this essay.

8. Of course, one need bear in mind that simulation of stochastic dynamics by digital machinery is always imperfect.

9. Bruce H. Weber, personal communication.

10. John Fellows, personal communication.

11. Irreversibility in the evolutionary trajectory is readily demonstrated, as recent work with Darwin's famous finches has shown. Peter Grant (1999; see also Jørgensen and Fath 2004) studied near-term changes in the beak sizes of Darwin's finches on the Galapagos Islands. He and wife Rosemary observed that, as drought made smaller, softer seeds less available to finches, selection began to favor larger beaks. The reverse was not observed, however. When rainfall returned to normal, beak size did not follow suit because smaller, softer seeds could still be processed by large beaks.

Chapter 3: How Can Things Truly Change?

1. Actually, Elsasser used 10^{85}, but more recent estimates put the number nearer to 10^{81}. In any event, the precise value of this exponent is not critical to Elsasser's argument.

2. 75! (*read* seventy-five factorial) signifies the product of all the integers up to and including 75. That is, 75 x 74 x 73 x 72 x . . . x 1.

3. Ecologist Sven Jørgensen (1995) estimated the number of observations it would take to completely specify a far simpler ecosystem than this, and he calculated a number that was enormous in the sense of Elsasser.

Chapter 4: How Can Things Persist?

1. I remind the reader of the operational definition of process adopted in chapter 2: "A process is the interaction of random events upon a configuration of constraints that results in a nonrandom but indeterminant outcome."

2. Actually, much of what is treated under the rubric of autocatalysis does not involve true catalysis. A catalyst is something that speeds up a reaction without itself being changed. The phenomenon under discussion might more accurately be termed *autofacilitation* because that which accelerates is itself changed (Terrance Deacon, personal communication).

The term *autocatalysis* will, nonetheless, be retained, as the term is now broadly understood to encompass autofacilitation.

3. Alicia Juarrero, personal communication.

4. A. Juarrero, personal communication.

5. Brooke Parry Hecht, personal communication.

6. James Coffman, personal communication.

7. Peter M. Allen, personal communication.

8. An example calculation similar to that used to estimate the AMI will be given in the next chapter.

9. Actually, my method of scaling was not motivated by Bateson's considerations, about which I was unaware at the time. I had simply drawn upon my background in thermodynamics to define ascendency in analogy to the Gibbs or Helmholtz free energies (measures of the capacity of a system to do effective work [Schroeder 2000]). Both of these quantities take the form of a scalar measure of the system's energy multiplied by a logarithmic term (as with the AMI) indicative of its constitution.

Chapter 5: Agency in Evolutionary Systems

1. Because *dialectic* refers strictly to human interactions and because the term often carries undesirable socio/political baggage, I prefer the adjective *transactional*. The two terms will be used interchangeably.

2. By *variational* is meant that the problem is cast as the optimization of some single objective function. For example, one might seek to maximize the profit on a particular economic venture.

3. The term *catastrophic* is being used here in the mathematical sense, meaning a jump discontinuity. The usual negative connotation does not necessarily apply.

4. Polymers are molecular chains of smaller molecular building blocks called monomers.

5. Philip Clayton, personal communication.

6. J. Coffman, personal communication.

7. This definition of information as a *decrease in indeterminacy* is an important clarification. All too many treatments of information theory muddy the waters by conflating information with indeterminacy itself. See chapter 5 in Ulanowicz (1986).

8. *Equilibrium* in thermodynamics has a very narrow meaning. If a system is suddenly cut off from its environment and it does not change after isolation, then it was originally in equilibrium.

Chapter 6: An Ecological Metaphysic

1. P. Clayton, personal communication.

2. Emilio Del Giudice, personal communication.

3. Daniel Brooks, personal communication.

4. Eddington's original reproach read, "If someone points out to you that your pet theory of the universe is in disagreement with Maxwell's equations—then so much the worse for Maxwell's equations. If it is found to be contradicted by observation, well, these experimentalists do bungle things sometimes. But if your theory is found to be against the second law of thermodynamics I can give you no hope; there is nothing for it but to collapse in deepest humiliation."

5. A. Juarrero and Bernard Patten, personal communications.

Chapter 7: The View out the Window

1. Jerome L. Chandler, personal communication.

2. It is curious that Newton himself saw no inconsistency on God's part in breaking any of the laws that Newton himself was engaged in formulating (Davis 1990).

3. Physicists will also relate that the difference between object and process (e.g., particle and wave) becomes ambiguous at this level.

References

Allen, T. F. H., and T. B. Starr. 1982. *Hierarchy*. Chicago: University of Chicago Press.

Allesina, S., A. Bodini, and C. Bondavalli 2005. Ecological subsystems via graph theory: The role of strongly connected components. *Oikos* 110 (1): 164–76.

Allesina, S., and C. Bondavalli. 2004. WAND: An ecological network analysis user-friendly tool. *Environmental Modelling Software* 19 (4): 337–40.

Aquinas, T. 1981. *Summa theologica*. Westminster, MD: Christian Classics.

Atlan, H. 1974. On a formal definition of organization. *Journal of Theoretical Biology* 45: 295–304.

Baird, D., and R. E. Ulanowicz. 1989. The seasonal dynamics of the Chesapeake Bay ecosystem. *Ecolological Monographs* 59: 329–64.

Barabási, A-L. 2002. *Linked: The New Science of Networks*. Cambridge, MA: Perseus.

Bateson, G. 1972. *Steps to an Ecology of Mind*. New York: Ballantine Books.

Bateson, G., and M. C. Bateson. 1987. *Angels Fear: Towards an Epistemology of the Sacred*. New York: Macmillan.

Behe, M. J. 1996. *Darwin's Black Box: The Biochemical Challenge to Evolution*. New York: Free Press.

Bersier, L. 2002. Quantitative descriptors of food web matrices. *Ecology* 83 (9): 2394–2407.

Bickhardt, M. H., and D. T. Campbell. 1999. Emergence. In P. B. Andersen, C. Emmeche, N. O. Finnemann, and P. V. Christiansen, eds., *Downward Causation*, 322–48. Aarhus, DK: Aarhus University Press.

Bohm, D. 1989. *Quantum Theory*. Mineola, NY: Dover.

Bohr, N. 1928. The quantum postulate and the recent development of atomic theory. *Nature* 121: 580–91.

Bondavalli, C., and R. E. Ulanowicz. 1999. Unexpected effects of predators upon their prey: The case of the American alligator. *Ecosystems* 2: 49–63.

Borgatti, S. P., and P. Foster. 2003. The network paradigm in organizational research: A review and typology. *Journal of Management* 29 (6): 991–1013.

Bose, P., D. H. Albonesi, and D. Marculescu. 2003. Power and complexity aware design. *IEEE Micro* 23 (5): 8–11.

Bosserman, R. W. 1981. Sensitivity techniques for examination of input-output flow analyses. In W. J. Mitsch and J. M. Klopatek, eds., *Energy and Ecological Modelling*, 653–60. Amsterdam: Elsevier.

Brinson, M. M., R. R. Christian, and L. K. Blum. 1995. Multiple states in the sea-level induced transition from terrestrial forest to estuary. *Estuaries* 18: 648–59.

Brooke, J. H. Forthcoming. Should the word "nature" be eliminated? In J. Proctor, ed. *New Visions in Nature, Science and Religion*. West Conshohocken, PA: Templeton Foundation Press.

Brooks, D. R., and E. O. Wiley. 1988. *Evolution as Entropy: Toward a Unified Theory of Biology*. 2nd ed. Chicago: University of Chicago Press.

Buffon, G. L. L. 1778. *Histoire naturelle, général et particulièr, Supplement*. Vol. 5. Paris: De l'imprimerie royale.

Callahan, S. 2003. *Women Who Hear Voices*. New York: Paulist Press.

Carathéodory, C. 1909. Investigation into the foundations of thermodynamics. *Mathematische Annalen* (Berlin) 67: 355–86.

Carnot, S. 1824/1943. *Reflections on the Motive Power of Heat*. Trans. R. H. Thurston. New York: ASME.

Chaisson, E. J. 2001. *Cosmic Evolution: The Rise of Complexity in Nature*. Cambridge, MA: Harvard University Press.

Chardin, T. de. 1966. *Man's Place in Nature*. New York: Harper & Row.

Christensen, V., and D. Pauly. 1992. ECOPATH II—A software for balancing steady-state models and calculating network characteristics. *Ecological Modelling* 61: 169–85.

Churchland, P. M., and P. S. Churchland. 1998. *On the Contrary: Critical Essays, 1987–1997*. Cambridge, MA: MIT Press.

Clayton, Philip. 2004. *Mind and Emergence: From Quantum to Consciousness*. New York: Oxford University Press.

Clements, F. E. 1916. *Plant Succession: An Analysis of the Development of Vegetation*. Washington, D.C.: Carnegie Institution of Washington, D.C.

Cody, G. D., R. M. Hazen, J. A. Brandes, H. Morowitz, and H. S. Yoder, Jr. 2001. The geochemical roots of Archeic autrotrophic carbon fixation: Hydrothermal experiments in the system Citric-H2O +/-FeS +/-NiS. *Geochimica et Cosmochimica Acta* 65 (20): 3557–76.

Coffman, J. A. 2006. Developmental ascendency: From bottom-up to top-down control. *Biological Theory* 1 (2): 165–78.

Cohen, J. E. 1976. Irreproducible results and the breeding of pigs. *Bioscience* 26: 391–94.

Cohen, J. E., F. Briand, and C. M. Newman. 1990. *Community Food Webs: Data and Theory*. Biomathematics 20. New York: Springer-Verlag.

Cohen, J. E., R. Beaver, S. Cousins, D. DeAngelis, L. Goldwasser, K. L. Heong, R. Holt, A. Kohn, J. Lawton, N. Martinez, R. O'Malley, R. Page, B. Patten, S. Pimm, G. Polis, M. Rejmanek, T. Schoener, K. Schoenly, W. R. Spurles, J. Teal, R E. Ulanowicz, P. Warren, H. Wilbur, and P. Yodzis. 1993. Improving food webs. *Ecology* 74: 252–58.

Collier, J. D. 1990. Intrinsic information. In P. P. Hanson, ed., *Information, Language and Cognition*, 390–409. Vancouver Studies in Cognitive Science, vol. 1. Vancouver, BC: British Columbia Press.

Corliss, J. B. 1992. The submarine hot spring hypothesis for the origin of life on earth. http://www.ictp.trieste.it/~chelaf/postdeadline.html.

Cuvier, G. 1825/1969. *Discours sur les révolutions de la surface du globe, et sur les changements qu'elles ont produits dans le règne animal*. 3rd ed. Bruxelles: Culture et civilisation.

Crick, F. H. C., and L. E. Orgel. 1973. Directed panspermia. *Icarus* 19: 341–46.

Daly, H. E. 2004. Reconciling the economics of social equity and environmental sustainability. *Population and Environment* 24 (1): 47–53.

Darwin, C. 1859. On the origin of the species by means of natural selection, or, the preservation of favoured races in the struggle for life. London: J. Murray.

Davidson, E. H. 2006. *The Regulatory Genome: Gene Regulatory Networks in Development and Evolution*. Philadelphia: Elsevier.

Davies, P. C. W., and J. R. Brown. 1986. The strange world of the quantum. In P. C. W. Davies and J. R. Brown, eds., *The Ghost in the Atom: A Discussion of the Mysteries of Quantum Physics*, 1–39. Cambridge: Cambridge University Press.

Davis, E. B. 1990. Newton's rejection of the "Newtonian world view": The role of divine will in Newton's natural philosophy. *Fides et Historia* 22: 6–20.

———. 1998. A God who does not itemize versus a science of the sacred. *American Scientist* 86: 572–74.

Dawkins, R. 1976. *The Selfish Gene*. New York: Oxford University Press.

———. 2006. *The God Delusion*. New York: Bantam.

Deacon, T. W. 2006. Reciprocal linkage between self-organizing processes is sufficient for self-reproduction and evolvability. *Biological Theory* 1 (2): 136–49.

DeAngelis, D. L., W. M. Post, and C. C. Travis. 1986. *Positive Feedback in Natural Systems*. New York: Springer-Verlang.

D'Elia, C. F. 1988. Nitrogen versus phosphorous. In S. E. McCoy, ed., *Chesapeake Bay: Issues, Resources, Status and Management*, 69–87. NOAA Estuary-of-the-Month Seminar No. 5. Washington, D.C.: U.S. Department of Commerce.

Dembski, W. A. 1998. Intelligent design as a theory of information. *Access Research Network*. http://www.arn.org/docs/dembski/wd_idtheory.htm.

Dennett, D. C. 1995. *Darwin's Dangerous Idea: Evolution and the Meanings of Life.* New York: Simon and Schuster.

———. 2006. *Breaking the Spell: Religion as a Natural Phenomenon.* New York: Viking.

Depew, D. J., and B. H. Weber. 1995. *Darwinism Evolving: Systems Dynamics and the Geneology of Natural Selection.* Cambridge, MA: MIT Press.

Eddington, A. E. 1930. *The Nature of the Physical World.* New York: Mac-Millan.

Eigen, M. 1971. Selforganization of matter and the evolution of biological macromolecules. *Naturwiss* 58: 465–523.

Ellis, G. F. R. 2005. Physics, complexity and causality. *Nature* 435: 743.

Elsasser, W. M. 1969. A causal phenomena in physics and biology: A case for reconstruction. *American Scientist* 57: 502–16.

———. 1981. A form of logic suited for biology? In Robert Rosen, ed., *Progress in Theoretical Biology*, 6: 23–62. New York: Academic Press.

Fath, B. D., and B. C. Patten. 1998. Network synergism: Emergence of positive relations in ecological models. *Ecological Modelling* 107: 127–43.

———. 1999. Review of the foundations of network environ analysis. *Ecosystems* 2: 167–79.

Feyerabend, P. K. 1978. *Against Method: Outline of an Anarchistic Theory of Knowledge.* New York: Verso.

Finn, J. T. 1976. Measures of ecosystem structure and function derived from analysis of flows. *Journal of Theoretical Biology* 56: 363–80.

Fiscus, D. A. 2001. The ecosystemic life hypothesis I: Introduction and definitions. *Bulletin of the Ecological Society of America* 82 (4): 248–50.

Fox, S. W. 1995. Thermal synthesis of amino acids and the origin of life. *Geochimica et Cosmochimica Acta* 59: 1213–14.

Fox-Keller, E. 2007. A clash of two cultures. *Nature* 445: 603.

Franklin, J. 1983. *Not Quite a Miracle: Brain Surgeons and Their Patients on the Frontier of Medicine.* Garden City, NY: Doubleday.

Fukuyama, F. 2002. *The End of History and the Last Man.* New York: Perennial.

Gibson, J. J. 1979. *The Ecological Approach to Visual Perception.* Boston: Houghton Mifflin.

Gladyshev, G. P. 1997. *Thermodynamic Theory of the Evolution of Living Beings.* Commack, NY: Nova Scientific Publishers.

Gattie D. K., E. W. Tollner, and T. L. Foutz. 2005. Network environ analysis: A mathematical basis for describing indirect effects in ecosystems. *Transactions of ASAE* 48 (4): 1645–52.

Gleason, H. A. 1917. The structure and development of the plant association. *Bulletin Torrey Botanical Club* 44: 463–81.

Gödel, K. 1931. Über formal unentscheidbare Sätze der *Principia Mathematica* und verwandter Systeme. *Monatshefte für Mathematik und Physik* 38: 173–98.

Goerner, S. J. 1999. *After the Clockwork Universe: The Emerging Science and Culture of Integral Society*. Edinburgh: Floris Books.

Goethe, J. W. von. 1775/1958. *Urfaust*. Ed. R. H. Samuel. London: Macmillan.

Gould, S. J. 2002. *The Structure of Evolutionary Theory*. Cambridge, MA: Belknap Press.

Gould, S. J., and N. Eldredge. 1977. Punctuated equilibria: The tempo and mode of evolution reconsidered. *Paleobiology* 3: 115–51.

Grant. P. R. 1999. *Ecology and Evolution of Darwin's Finches*. 2nd ed. Princeton, NJ: Princeton Science Library.

Grenz, S. J. 1996. *A Primer on Postmodernism*. Grand Rapids, MI: William B. Eerdmans.

Griffin, D. R. 1996. Introduction to SUNY series in constructive postmodern thought. In F. Ferre, *Being and Value*, xv–xviii. Albany, NY: SUNY Press.

Guttalu, R. S., and H. Flashner. 1989. Periodic solutions of non-linear autonomous systems by approximate point mappings. *Journal of Sound and Vibration* 129 (2): 291–311.

Hagen, J. B. 1992. *An Entangled Bank: The Origins of Ecosystem Ecology*. New Brunswick, NJ: Rutgers University Press.

Haken, H., 1988. *Information and Self-Organization: A Macroscopic Approach to Complex Systems*. Berlin: Springer-Verlag.

Hallsdottir, M. 1995. On the pre-settlement history of Icelandic vegetation. *Icelandic Agricultural Sciences* 9: 17–29.

Hannon, B. 1973. The structure of ecosystems. *Journal of Theoretical Biology* 41: 535–46.

Haught, J. F. 2000. *God after Darwin: A Theology of Evolution*. Boulder, CO: Westview Press.

———. 2001. Science, religion, and the origin of life. Paper delivered at the AAAS Seminar, Washington, D.C. September 13.

———. 2003. *Deeper than Darwin: The Prospect for Religion in the Age of Evolution*. Boulder, CO: Westview Press.

Hawking, S. W. 1988. *A Brief History of Time: From the Big Bang to Black Holes*. New York: Bantam.

Hefner, P. 2000. Why I don't believe in miracles. *Newsweek* 135(17): 61 (May 1).

References

Heisenberg, W. 1927. Ueber den anschaulichen Inhalt der quantentheoretischen Kinematik und Mechanik. *Zeits. f. Physik* 43: 172–98.

Hirata, H., and R. E. Ulanowicz. 1984. Information theoretical analysis of ecological networks. *International Journal of Systems Science* 15: 261–70.

Ho, M-W. 1993. *The Rainbow and the Worm: The Physics of Organisms.* River Edge, NJ: World Scientific.

Ho, M-W., and R. E. Ulanowicz. 2006. Sustainable systems as organisms? *BioSystems* 82: 39–51.

Hodge, M. J. S. 1992. Biology and philosophy (including ideology): A study of Fisher and Wright. In S. Sarkar, ed., *The Founders of Evolutionary Genetics*, 231–93. Doderecht: Kluwer.

Hoffmeyer, J. 2008. *Biosemiotics: Signs of Life and Life of Signs.* Scranton, PA: University of Scranton Press.

Holling, C. S. 1978. The spruce-budworm/forest-management problem. In C. S. Holling, ed., *Adaptive Environmental Assessment and Management*, 143–82. International Series on Applied Systems Analysis 3. Hoboken, NJ: John Wiley & Sons.

———. 1986. The resilience of terrestrial ecosystems: Local surprise and global change. In W. C. Clark and R. E. Munn, eds., *Sustainable Development of the Biosphere*, 292–317. Cambridge: Cambridge University Press.

Huberman, B. A., ed. 1988. *The Ecology of Computation.* Amsterdam, NY: North-Holland.

Hutchinson, G. E. 1948. Circular causal systems in ecology. *Annals of the New York Academy of Sciences* 50: 221–46.

Jackelén, A. Forthcoming. Creativity through emergence: A vision of nature and God. In J. Proctor, ed., *New Visions in Nature, Science and Religion.* West Conshohocken, PA: Templeton Foundation Press.

Jolly, A. 2006. A global vision. *Nature* 443: 148.

Jonas, H. 1966. *The Phenomenon of Life.* New York: Harper and Row.

Jørgensen, S. E. 1992. *Integration of System Theories: A Pattern.* Dordrecht: Kluwer.

———. 1995. Quantum mechanics and complex ecology. In B. C. Patten and S. E. Jørgensen, eds., *Complex Ecology: The Part–Whole Relation in Ecosystems*, 34–39. Englewood Cliffs, NJ: Prentice Hall.

Jørgensen, S. E., and B. D. Fath. 2004. Modelling the selective adaptation of Darwin's finches. *Ecological Modelling* 176 (3–4): 409–18.

Jørgensen, S. E. and H. Mejer. 1979. A holistic approach to ecological modelling. *Ecological Modelling* 7: 169–89.

Jørgensen, S. E., B. D. Fath, S. Bastianoni, J. Marques, F. Mueller, S. Nors-Nielsen, B. C. Patten, E. Tiezzi, and R. E. Ulanowicz. 2007. *A New Ecology: Systems Perspective.* Amsterdam: Elsevier.

Juarrero, A. 1999. *Dynamics in Action: Intentional Behavior as a Complex System*. Cambridge, MA: MIT Press.

Kauffman, S. 1995. *At Home in the Universe: The Search for the Laws of Self-Organization and Complexity*. New York: Oxford University Press.

———. 2000. *Investigations*. Oxford: Oxford University Press.

Keller, C. 2005. *God and Power: Counter-Apocalyptic Journeys*. Minneapolis, MN: Fortress Press.

Kikawada, H. 1998. Applying network analysis to simulated ecological landscape dynamics. Master's thesis, University of Maryland, College Park, MD.

Kolasa, J., and S. T. A. Pickett. 1991. *Ecological Heterogeneity*. New York: Springer-Verlag.

Konopka, A. K. 2007. Basic concepts of systems biology. In A. K. Konopka, ed., *Systems Biology: Principles, Methods, and Concepts*, 1–26. Boca Raton, FL: CRC Press.

Krause, A. E., K. A. Frank, D. M. Mason, R. E. Ulanowicz, and W. W. Taylor. 2003. Compartments revealed in food-web structure. *Nature* 426: 282–85.

Kristinsson, H. 1995. Post-settlement history of Icelandic forests. *Icelandic Agricultural Sciences* 9: 31–35.

Krivov, S., and R. E. Ulanowicz. 2003. Qunatitative measures of organization for multi-agent systems. *BioSystems* 69: 39–54.

Laplace, P. S. 1814/1951. *A Philosophical Essay on Probabilities*. Trans. F. W. Truscott and F. L. Emory. New York: Dover Publications, Inc.

Leontief, W. 1951. *The Structure of the American Economy, 1919–1939*. 2nd ed. New York: Oxford University Press.

Levine, S. 1980. Several measures of trophic structure applicable to complex food webs. *Journal of Theoretical Biology* 83: 195–207.

Lewontin, R. C. 1997. Billions and billions of demons. *The New York Review of Books* 44(1): 28–32 (January 9).

———. 2000. *The Triple Helix: Gene Organism and Environment*. Cambridge, MA: Harvard University Press.

Liljenstroem, H., and U. Svedin, eds. 2005. *Micro, Meso, Macro: Addressing Complex Systems Couplings*. Hackensack, NJ: World Scientific.

Lindeman, R. L. 1942. The trophic-dynamic aspect of ecology. *Ecology* 23: 399–418.

Long, R. R. 1963. *Engineering Science Mechanics*. Englewood Cliffs, NJ: Prentice Hall.

Lorenz, R. D. 2002. Planets, life and the production of entropy. *International Journal of Astrobiology* 1: 3–13.

Lotka, A. J. 1922. Contribution to the energetics of evolution. *Proceedings of the Natural Academy of Sciences* 8: 147–50.

References

Lyell, C. 1830. *Principles of geology, being an attempt to explain the former changes of the earth's surface, by reference to causes now in operation.* London: J Murray.

MacArthur, R. H. 1955. Fluctuations of animal populations and a measure of community stability. *Ecology* 36: 533–36.

Mason, D. M. 2006. *Econetwrk: A Windows-compatible Tool to Analyze Ecological Flow Networks.* Ann Arbor, MI: NOAA Great Lakes Environmental Laboratory. http://www.glerl.noaa.gov/EcoNetwrk/.

Matsuno, K. 1996. Internalist stance and the physics of information. *Journal of Biological Systems* 38: 111–18.

Maturana, H. R., and F. J. Varela. 1980. *Autopoiesis and Cognition: The Realization of the Living.* Dordrecht: D. Reidel.

Matutinović, I. 2002. Organizational patterns of economies: An ecological perspective. *Ecological Economics* 3: 421–40.

———. 2005. The microeconomic foundations of business cycles: From institutions to autocatalytic networks. *Journal of Economic Issues* 39 (4): 867–98.

May, R. M. 1972. Will a large complex system be stable? *Nature* 238: 413–14.

———. 1981. *Theoretical Ecology: Principles and Applications.* Sunderland, MA: Sinauer.

Mayr, E. 1997. *This Is Biology: The Science of the Living World.* Cambridge, MA: The Belknap Press of Harvard University Press.

McLuhan, M. 1964. *Understanding Media.* New York: Mentor.

Mickulecky, D. C. 1985. Network thermodynamics in biology and ecology: An introduction. In R. E. Ulanowicz and T. Platt, eds., *Ecosystem Theory for Biological Oceanography*, 163–75. Canadian Bulletin of Fisheries and Aquatic Sciences 213, Ottawa.

Miller, S. L., and H. C. Urey. 1959. Organic compound synthesis on the primitive earth. *Science* 130: 245.

Morrison, P., J. Billingham, and J. Wolfe, eds. 1977. *The Search for Extraterrestrial Intelligence: SETI.* Washington, D.C.: NASA.

Mueller, F., and M. Leupelt. 1998. *Eco Targets, Goal Functions, and Orientors.* Berlin: Springer-Verlag.

Murphy, N., and W. S. Brown. 2007. *Did My Neurons Make Me Do It? Philosophical and Neurobiological Perspectives on Moral Responsibility and Free Will.* Oxford: Oxford University Press.

Naess, A. 1988. Deep ecology and ultimate premises. *Ecologist* 18: 128–31.

Noether, A. 1983. *Gesammelte Abhandlungen.* Ed. N. Jaconson. New York: Springer Verlag.

Odum, E. P. 1953. *Fundamentals of Ecology.* Philadelphia: Saunders.

———. 1969. The strategy of ecosystem development. *Science* 164: 262–70.

———. 1977. The emergence of ecology as a new integrative discipline. *Science* 195: 1289–93.

Odum, E. P., and H. T. Odum. 1959. *Fundamentals of Ecology*. 2nd ed. Philadelphia: Saunders.

Odum, H. T. 1971. *Environment, Power and Society*. New York: John Wiley and Sons.

Odum, H. T., and R. C. Pinkerton. 1955. Time's speed regulator: The optimum efficiency for maximum power output in physical and biological systems. *American Scientist* 43: 331–43.

Pahl-Wostl, C. 1995. *The Dynamic Nature of Ecosystems: Chaos and Order Entwined*. New York: Wiley.

Patten, B. C. 1999. Out of the clockworks. *Estuaries* 22 (2A): 339–42.

Penrose, O. 2005. An asymmetric world. *Nature* 438: 919.

Penrose, R. 1994. *Shadows of the Mind*. Oxford: Oxford University Press.

———. *The Road to Reality: A Complete Guide to the Laws of the Universe*. New York: A. A. Knopf.

Peterson, G. R. Forthcoming. Is nature radically emergent? In J. Proctor, ed., *New Visions in Nature, Science and Religion*. West Conshohocken, PA: Templeton Foundation Press.

Peirce, C. S. 1892. The doctrine of necessity examined. *The Monist* 2: 321–38.

Pielou, E. C. 1975. *Ecological Diversity*. New York: Wiley.

Platt, T., K. H. Mann, and R. E. Ulanowicz, eds. 1981. *Mathematical Models in Biological Oceanography*. Paris: UNESCO Press.

Popper, K. R. 1959. *The Logic of Scientific Discovery*. London: Hutchinson.

———. 1982. *The Open Universe: An Argument for Indeterminism*. Totowa, NJ: Rowman and Littlefield.

———. 1990. *A World of Propensities*. Bristol, UK: Thoemmes.

Prigogine, I. 1945. Moderation et transformations irreversibles des systemes ouverts. *Bulletin de la classe des sciences—Académie royale de Belgique*. 31: 600–606.

———. 1978. Time, structure and fluctuations. *Science* 201: 777–85.

Prigogine, I., and I. Stengers. 1984. *Order out of Chaos: Man's New Dialogue with Nature*. New York: Bantam.

Rabin, B. S. 1999. *Stress, Immune Function, and Health: The Connection*. New York: Wiley-Liss.

Rosen, R. 1985a. Information and complexity. In R. E. Ulanowicz and T. Platt, eds., *Ecosystem Theory for Biological Oceanography*, 221–33. Canadian Bulletin of Fisheries and Aquatic Sciences 213. Ottawa.

———. 1985b. *Anticipatory Systems: Philosophical, Mathematical, and Methodological Foundations*. Oxford: Pergamon Press.

Rosen, R. 2000. *Essays on Life Itself*. New York: Columbia University Press.

Russell, B. 1993. *An Outline of Philosophy*. New York, NY: Routledge.

Rutledge, R. W., B. L. Basorre, and R. J. Mulholland. 1976. Ecological stability: An information theory viewpoint. *Journal of Theoretical Biology* 57: 355–71.

Salthe, S. N. 1989. Commentary on "The influence of the evolutionary paradigms" by R. M. Burian. In M. K. Hecht, ed., *Evolutionary Biology at the Crossroads: A Symposium at Queens College*, 174–76. Flushing, NY: Queens College Press.

———. 1993. *Development and Evolution: Complexity and Change in Biology*. Cambridge, MA: MIT Press.

Schneider, E. D., and J. J. Kay. 1994. Life as a manifestation of the second law of thermodynamics. *Mathematical and Computer Modelling* 19: 25–48.

Schneider, E. D., and D. Sagan. 2005. *Into the Cool: Energy Flow, Thermodynamics and Life*. Chicago: University of Chicago Press.

Schroeder, D. V. 2000. *An Introduction to Thermal Physics*. San Francisco: Addison Wesley.

Schroedinger, E. 1944. *What Is Life?* Cambridge: Cambridge University Press.

Shannon, C. E. 1948. A mathematical theory of communication. *Bell System Technology Journal* 27: 379–423.

Smith, J. M. 1982. *Evolution and the Theory of Games*. Cambridge: Cambridge University Press.

Snow, C. P. 1963. *The Two Cultures and the Scientific Revolu*tion. Cambridge: Cambridge University Press.

Solzhenitsyn, A. I. 1989. *August 1914*. New York: Farrar, Straus and Giroux.

Sonnenschein, C., and A. M. Soto. 2006. Emergentism by default: A view from the bench. *Synthese* 151 (3): 361–76.

Strohman, R. C. 2001. Human genome project in crisis: Where is the program for life? *California Monthly* 111 (5): 24–27.

Susskind, L. 2005. The landscape. http://www.edge.org/q2006/q06_6 .html.

Szyrmer, J., and R. E. Ulanowicz. 1987. Total flows in ecosystems. *Ecological Modelling* 35: 123–36.

Tiezzi, E. 2006. *Steps towards an Evolutionary Physics*. Southampton, UK: WIT Press.

Tilly, L. J. 1968. The structure and dynamics of Cone Spring. *Ecological Monographs* 38: 169–97.

Tribus, M., and E. C. McIrvine. 1971. Energy and information. *Scientific American* 225 (3): 179–88.

Ulanowicz, R. E. 1980. An hypothesis on the development of natural communities. *Journal of Theoretical Biology* 85: 223–45.

———. 1983. Identifying the structure of cycling in ecosystems. *Mathematical Biosciences* 65: 219–37.

———. *Growth & Development: Ecosystems Phenomenology*. New York: Springer-Verlag.

———. 1995a. Beyond the material and the mechanical: Occam's razor is a double-edged blade. *Zygon* 30 (2): 249–66.

———. 1995b. Utricularia's secret: The advantage of positive feedback in oligotrophic environments. *Ecological Modelling* 79: 49–57.

———. 1995c. Ecosystem trophic foundations: Lindeman Exonerata. In B. C. Patten and S. E. Jørgensen, eds. *Complex Ecology: The Part–Whole Relation in Ecosystems*, 549–60. Englewood Cliffs, NJ: Prentice Hall.

———. 1997. *Ecology, the Ascendent Perspective*. New York: Columbia University Press.

———. 1999a. Life after Newton: An ecological metaphysic. *Biological Systems* 50: 127–42.

———. 1999b. Out of the clockworks: A response. *Estuaries* 22: 342–43.

———. 2001. The organic in ecology. *Ludus Vitales* 9 (15): 183–204.

———. 2002. Ecology, a dialogue between the quick and the dead. *Emergence* 4: 34–52.

———. 2004a. New perspectives through brackish water ecology. *Hydrobiologia* 514: 3–12.

———. 2004b. Quantitative methods for ecological network analysis. *Computational Biology and Chemistry* 28: 321–39.

———. 2004c. Ecosystem dynamics: A natural middle. *Theology and Science* 2 (2): 231–53.

———. 2005. Fluctuations and order in ecosystem dynamics. *Emergence: Complexity and Organization* 7 (2): 14–20.

———. 2006. Process ecology: A transactional worldview. *Journal of Ecodynamics* 1: 103–14.

———. 2007. Emergence, naturally! *Zygon* 42 (4): 945–60.

———. Forthcoming. Enduring metaphysical impatience? In J. Proctor, ed., *New Visions in Nature, Science and Religion*. West Conshohocken, PA: Templeton Foundation Press.

Ulanowicz, R. E., and D. Baird. 1999. Nutrient controls on ecosystem dynamics: The Chesapeake mesohaline community. *Journal of Marine Systems* 19: 159–72.

Ulanowicz, R. E., and B. M. Hannon. 1987. Life and the production of entropy. *Proceedings of the Royal Society, London B.* 232: 181–92.

Ulanowicz, R. E., and J. J. Kay. 1991. A package for the analysis of ecosystem flow networks. *Environmental Software* 6: 131–42.

Ulanowicz, R. E., and J. S. Norden. 1990. Symmetrical overhead in flow networks. *International Journal of Systems Science* 1: 429–37.

Ulanowicz, R. E., and C. J. Puccia. 1990. Mixed trophic impacts in ecosystems. *Coenoses* 5: 7–16.

Ulanowicz, R. E., and M. J. Zickel. 2005. Using ecology to quantify organization in fluid flows. In A. Kleidon and R. Lorenz, eds., *Non-*

Equilibrium Thermodynamics and the Production of Entropy: Life, Earth, and Beyond, 57–66. Berlin: Springer Verlag.

Vanderburg, W. H. 1990. Integrality, context, and other industrial casualties. In C. J. Edwards and H. A. Regier, eds., *An Ecosystem Approach to the Integrity of the Great Lakes in Turbulent Times*, 79–90. Great Lakes Fishery Commission, Special Publication 90-4. Ann Arbor, MI.

Vaz, N. M., and C. R. Carvalho. 1994. Assimilation, tolerance and the end of innocence. *Ciência e Cultura* (São Paulo) 46 (5/6): 351–57.

Waddington, C. H. 1942. Canalisation of development and the inheritance of acquired characters. *Nature* 150: 563–65.

Watson, J. D., and F. H. C. Crick. 1953. Molecular structure of nucleic acids: A structure for deoxyribose nucleic acid. *Nature* 171: 737–38.

Westfall, R. S. 1983. *Never at Rest: A Biography of Isaac Newton*. Cambridge: Cambridge University Press.

———. 1993. *The Life of Isaac Newton*. Cambridge: Cambridge University Press.

Whitehead, A. N. 1929. *Process and Reality*. New York: Macmillan.

———. 1938. *Modes of Thought*. New York: The Free Press.

Whitehead, A. N., and B. Russell. 1913. *Principia Mathematica*. Cambridge: Cambridge University Press.

Wicken, J. S., and R. E. Ulanowicz. 1988. On quantifying hierarchical connections in ecology. *Journal of Social and Biological Structures* 11: 369–77.

Williams, R. J. 1956. *Biochemical Individuality: The Basis for the Genetotrophic Concept*. Austin: University of Texas Press.

Wills, G. 1978. *Inventing America*. Garden City, NY: Doubleday.

Wilson, E. O. 1980. *Sociobiology*. Cambridge, MA: Harvard University Press.

———. 1998. *Consilience: The Unity of Knowledge*. New York: Knopf.

Wimsatt, W. C. 1995. The ontology of complex systems: Levels of organization, perspectives and causal thickets. *Canadian Journal of Philosophy* 20: 207–74.

Wolfram, S. 2002. *A New Kind of Science*. Champaign, IL: Wolfram Media.

Wright, R. 1988. Did the universe just happen? *Atlantic Monthly*, April, 29–44.

Zorach, A. C., and R. E. Ulanowicz. 2003. Quantifying the complexity of flow networks: How many roles are there? *Complexity* 8 (3): 68–76.

Zotin, A. A., and A. I. Zotin. 1997. Phenomenological theory of ontogenesis. *International Journal of Developmental Biology* 41: 917–21.

Name Index

Subject Index